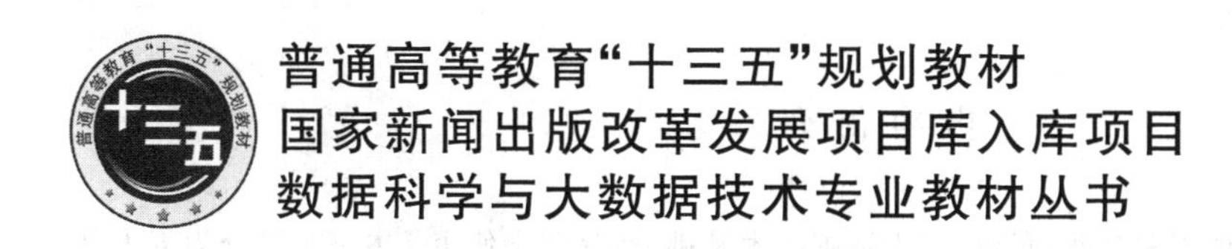

大数据技术基础实验

欧中洪　宋美娜　鄂海红　编著

北京邮电大学出版社
www.buptpress.com

内容简介

本书为“大数据技术基础”的实验教程，可以和《大数据技术基础》教材配套使用。本书主要分为五大部分：大数据存储技术实验教程（分布式文件系统和 NoSQL 数据库），大数据处理框架实验教程（MapReduce 和 Spark、实时处理框架 Storm 和 Flink），大数据分析技术实验教程（Druid、Drill、Kylin），大数据可视化实验教程，大数据综合应用实验案例。本书围绕典型大数据应用系统所需的各个组成部分设计实验，指导读者开展大数据应用实践。本书可作为计算机学科相关专业，尤其是数据科学与大数据技术专业的专业教材，也可作为大数据相关专业从业人员的参考用书。

图书在版编目(CIP)数据

大数据技术基础实验 / 欧中洪，宋美娜，鄂海红编著. -- 北京：北京邮电大学出版社，2020.1

ISBN 978-7-5635-5873-5

Ⅰ. ①大… Ⅱ. ①欧… ②宋… ③鄂… Ⅲ. ①数据处理－实验－高等学校－教材 Ⅳ. ①TP274-33

中国版本图书馆 CIP 数据核字（2019）第 204846 号

书　　名：大数据技术基础实验

作　　者：欧中洪　宋美娜　鄂海红

责任编辑：徐振华　米文秋

出版发行：北京邮电大学出版社

社　　址：北京市海淀区西土城路 10 号(100876)

发 行 部：电话：010-62282185　传真：010-62283578

E-mail：publish@bupt.edu.cn

经　　销：各地新华书店

印　　刷：保定市中画美凯印刷有限公司

开　　本：787 mm×1 092 mm　1/16

印　　张：16

字　　数：419 千字

版　　次：2020 年 1 月第 1 版　2020 年 1 月第 1 次印刷

ISBN 978-7-5635-5873-5　　定价：48.00 元

·如有印装质量问题，请与北京邮电大学出版社发行部联系·

前　言

本书共分为8章。

第1章为大数据存储技术的实验教程，主要介绍主流分布式存储系统，包括HDFS常用操作、HBase的安装、HBase的Shell连接与数据操作、Redis数据库的安装和使用、MongoDB数据库的安装和使用、Neo4j数据库的安装和使用。

第2章主要介绍Hadoop框架中MapReduce的安装部署和基本操作，包括MapReduce的伪分布模式安装和完全分布模式安装，并以具体的单词计数、数据去重、二次排序、数据自定义格式输出等实验为例，利用MapReduce实现分布式并行计算。

第3章主要介绍基于内存的分布式计算框架Spark的安装部署和基本操作，包括Spark集群的安装部署，Spark-shell的基本操作，Spark Scala开发环境，PySpark的执行方式，Spark MLlib的基本使用方法等。

第4章主要介绍流处理框架的安装部署和基本操作，包括Storm的部署和基本操作，Flume数据收集系统的安装和使用，消息系统Kafka的安装和基本操作，Spark Streaming的安装和基本操作，Flink的安装和基本使用流程。

第5章主要介绍分布式查询引擎的安装部署和基本操作，包括Druid的安装部署、数据导入和查询，Drill的安装和使用。

第6章主要介绍大数据多维分析引擎的安装部署和使用，包括Kylin的安装和使用，创建多维分析Cube，利用REST方式访问Kylin。

第7章主要介绍大数据可视化系统的安装部署和基本操作，包括ECharts系统的安装和基本使用流程、利用ECharts绘制可视化图表等，以及Plotly可视化系统的安装和使用。

第8章为大数据综合实验，包括数据采集、预处理、存储、查询和可视化分析、数据挖掘等，实现数据全流程操作。

本书可作为数据科学与大数据技术专业本科高年级专业课教材，也可作为研究生相关专业的参考资料。同时，本书可以和《大数据技术基础》教材配套使用。《大数据技术基础》从大数据基础理论和技术方面提供相关知识，有助于构建大数据技术的知识体系。

本书的编写得到了北京邮电大学 PCN&CAD 中心、教育部信息网络工程研究中心和北京邮电大学计算机学院数据科学与服务中心多名教师与研究生的支持，他们是：宋美娜、鄂海红、宋俊德、毕秋波、韩鹏昊、田川、孔慧慧、赵淑晨、吴金盛、温宇飞、万仁山、谭泽华、陈小康、韦帅丽、朱永波，在此一并表示感谢。

感谢国家重点研发计划项目“大数据征信及智能评估技术”“科技资源与服务集成技术”“京津冀协同创新综合科技服务发展模式及支撑技术研究”“基于大数据的科技咨询技术与服务平台研发”，国家科技条件平台计划项目“国家人类遗传资源共享服务平台北京创新中心建设”等项目对本书的大力支持。

由于作者水平有限，书中难免存在不足和错误之处，恳请广大读者批评指正。

欧中洪
北京邮电大学

目　录

第1章

大数据存储:分布式文件系统及 NoSQL 数据库实验教程

1.1 HDFS 常用操作

1. 实验目的

① 理解 Hadoop 分布式文件系统(HDFS)体系结构。

② 学会使用 HDFS 操作常用的 Shell 命令。

③ 学会编写读写 HDFS 文件的 Java 程序。

2. 实验要求

① Linux、MacOSX 或其他类似于 Unix 的操作系统。

② JDK 1.8。

③ Hadoop 3.2.0。

3. 预备知识

HDFS 为大数据平台其他所有组件提供了最基本的存储功能,具有高容错性、高可靠性、可扩展、高吞吐率等特征,为大数据存储和处理提供了强大的底层存储架构。

HDFS 采用主/从体系结构模型,一个 HDFS 集群拥有一个 NameNode 和多个 DataNode。NameNode 管理文件系统的元数据,DataNode 存储实际的数据。客户端通过 NameNode 和 DataNode 的交互访问文件系统,联系 NameNode 以获取文件的元数据,而真正的文件 I/O 操作直接和 DataNode 进行交互[1]。

拓展阅读

HDFS 介绍

4. 实验内容

① 在实验前需先安装、配置 HDFS 及相应的依赖环境,建议使用单节点完成 HDFS 实验,具体安装配置可参考扩展资料。

② 进入 Hadoop sbin 文件夹,启动 HDFS 服务。

```
cd hadoop-3.2.0/sbin
./start-dfs.sh
```

③ 创建 user1 和 user2 目录，在 user1 目录中创建 remote. txt 文件，查看 user1 目录下的文件。

```
hadoop fs -mkdir /user1
hadoop fs -mkdir /user2
hadoop fs -touchz /user1/remote.txt
hadoop fs -ls /user1
```

④ 在本地创建 local. txt 文件，文件中写入“Hello，Hadoop！ Hello，HDFS!”，然后上传至 HDFS user2 目录。

```
hadoop fs -put local.txt /user2
```

⑤ 将 user1 目录下的 remote. txt 文件移动至 user2 目录下，查看源目录下的文件是否存在。

```
hadoop fs -mv /user1/remote.txt /user2
```

⑥ 将 user2 目录下的 local. txt 文件复制到 user1 目录下。

```
hadoop fs -cp /user2/local.txt /user1
```

⑦ 查看 user1 目录下的 local. txt 文件的内容。

```
hadoop fs -cat /user1/local.txt
```

⑧ 下载 user2 目录下的 remote. txt 文件到本地。

```
hadoop fs -get /user2/remote.txt
```

⑨ 删除 user2 目录下的 remote. txt 文件。

```
hadoop fs -rm /user2/remote.txt
```

⑩ 配置 Hadoop 的环境变量，编写 HDFS 读写的 Java 程序，编译程序并打成 jar 包，然后用 hadoop jar 命令运行。

5. 实验作业

① 参考扩展资料中的内容，完成 HDFS 的安装与配置。

② 按照实验内容的步骤，熟练使用 HDFS 操作的 Shell 命令。

③ 编写读写 HDFS 文件的 Java 程序。

6. 扩展资料

① 使用 HDFS 前需先搭建 Hadoop 环境。

Hadoop 官网下载地址：https://hadoop. apache. org/releases. html。

Hadoop 安装配置参考地址：https://hadoop. apache. org/docs/stable/hadoop-project-dist/hadoop-common/SingleCluster. html。

② 进入 Hadoop sbin 文件夹，启动 HDFS 服务，运行 jps 命令查看已启动服务，如图 1-1 所示。

```
root@slave2:/home/hadoop-3.2.0/sbin# ls
distribute-exclude.sh      start-all.sh              stop-balancer.sh
FederationStateStore       start-balancer.sh         stop-dfs.cmd
hadoop-daemon.sh           start-dfs.cmd             stop-dfs.sh
hadoop-daemons.sh          start-dfs.sh              stop-secure-dns.sh
httpfs.sh                  start-secure-dns.sh       stop-yarn.cmd
kms.sh                     start-yarn.cmd            stop-yarn.sh
mr-jobhistory-daemon.sh    start-yarn.sh             workers.sh
refresh-namenodes.sh       stop-all.cmd              yarn-daemon.sh
start-all.cmd              stop-all.sh               yarn-daemons.sh
root@slave2:/home/hadoop-3.2.0/sbin# ./start-dfs.sh
Starting namenodes on [localhost]
Starting datanodes
Starting secondary namenodes [slave2]
root@slave2:/home/hadoop-3.2.0/sbin# jps
4757 NameNode
5462 Jps
5308 SecondaryNameNode
4927 DataNode
```

图 1-1　HDFS 服务启动

③ 创建 user1 和 user2 目录，在 user1 目录中创建 remote. txt 文件，查看 user1 目录下的文件，如图 1-2 所示。

```
root@slave2:~# hadoop fs -mkdir /user1
root@slave2:~# hadoop fs -mkdir /user2
root@slave2:~# hadoop fs -touchz /user1/remote.txt
root@slave2:~# hadoop fs -ls /user1
Found 1 items
-rw-r--r--   1 root supergroup          0 2019-09-25 02:45 /user1/remote.txt
```

图 1-2　在 HDFS 上创建目录和文件

④ 在本地创建 local. txt 文件，文件中写入"Hello，Hadoop！ Hello，HDFS!"，然后上传至 HDFS 的 user2 目录，如图 1-3 所示。

```
root@slave2:~# vim local.txt
root@slave2:~# hadoop fs -put local.txt /user2
root@slave2:~# hadoop fs -ls /user2
Found 1 items
-rw-r--r--   1 root supergroup         28 2019-09-25 02:47 /user2/local.txt
```

图 1-3　上传 local. txt 文件到 HDFS 的 user2 目录

⑤ 将 user1 目录下的 remote. txt 文件移至 user2 目录下，查看源目录下的文件是否存在，如图 1-4 所示。

```
root@slave2:~# hadoop fs -mv /user1/remote.txt /user2
root@slave2:~# hadoop fs -ls /user1
root@slave2:~# hadoop fs -ls /user2
Found 2 items
-rw-r--r--   1 root supergroup         28 2019-09-25 02:47 /user2/local.txt
-rw-r--r--   1 root supergroup          0 2019-09-25 02:45 /user2/remote.txt
```

图 1-4　将 HDFS user1 目录下的 remote. txt 文件移动到 user2 目录下

⑥ 将 user2 目录下的 local. txt 文件复制到 user1 目录下，如图 1-5 所示。

```
root@slave2:~# hadoop fs -cp /user2/local.txt /user1
root@slave2:~# hadoop fs -ls /user1
Found 1 items
-rw-r--r--   1 root supergroup          28 2019-09-25 02:49 /user1/local.txt
root@slave2:~# hadoop fs -ls /user2
Found 2 items
-rw-r--r--   1 root supergroup          28 2019-09-25 02:47 /user2/local.txt
-rw-r--r--   1 root supergroup           0 2019-09-25 02:45 /user2/remote.txt
```

图 1-5　将 HDFS user2 目录下的 local. txt 文件复制到 user1 目录下

⑦ 查看 user1 目录下的 local. txt 文件的内容，如图 1-6 所示。

```
root@slave2:~# hadoop fs -cat /user1/local.txt
Hello,Hadoop!
Hello,HDFS!
```

图 1-6　查看 HDFS 上 local. txt 文件的内容

⑧ 下载 user2 目录下的 remote. txt 文件到本地，如图 1-7 所示。

```
root@slave2:~# hadoop fs -get /user2/remote.txt
root@slave2:~# ls
local.txt  remote.txt
```

图 1-7　下载 HDFS 上的 remote. txt 文件到本地

⑨ 删除 user2 目录下的 remote. txt 文件，如图 1-8 所示。

```
root@slave2:~# hadoop fs -rm /user2/remote.txt
Deleted /user2/remote.txt
root@slave2:~# hadoop fs -ls /user2
Found 1 items
-rw-r--r--   1 root supergroup          28 2019-09-25 02:47 /user2/local.txt
```

图 1-8　删除 HDFS 上的 remote. txt 文件

⑩ 添加 Hadoop 的环境变量，并使环境变量生效。

```
Export HADOOP_HOME = /home/hadoop - 3.2.0
Export CLASSPATH = $ CLASSPATH: $ HADOOP_HOME/share/hadoop/common/ * : $ HADOOP_
        HOME/share/hadoop/common/lib/ *
Export PATH = $ PATH: $ HADOOP_HOME/sbin: $ HADOOP_HOME/bin
Export HADOOP_COMMON_LIB_NATIVE_DIR = $ HADOOP_HOME/lib/native
Export HADOOP_OPTS = " - Djava. library. path = $ HADOOP_HOME/lib: $ HADOO P_HOME/
        lib/native"
```

⑪ 编译程序并打成 jar 包，如图 1-9 所示。

```
root@slave2:~# javac WriteFile.java
root@slave2:~# javac ReadFile.java
root@slave2:~# jar -cvf WriteFile.jar WriteFile.class
added manifest
adding: WriteFile.class(in = 1176) (out= 652)(deflated 44%)
root@slave2:~# jar -cvf ReadFile.jar ReadFile.class
added manifest
adding: ReadFile.class(in = 961) (out= 537)(deflated 44%)
root@slave2:~# ls
demo.txt   ReadFile.class  ReadFile.java  WriteFile.class  WriteFile.java
local.txt  ReadFile.jar    remote.txt     WriteFile.jar
```

图 1-9　编译 HDFS 读写程序并打成 jar 包

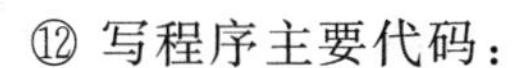

⑫ 写程序主要代码：

```
public class WriteFile {
  public static void main(String[] args) throws Exception {
    Configuration conf = new Configuration();
    FileSystem fileSystem = FileSystem.get(conf);
    String local = "demo.txt";
    FileInputStream inputStream = new FileInputStream(new File(local));
    String remote = "/user/demo.txt";
    FSDataOutputStream outputStream = fileSystem.create(new Path(remote));
    IOUtils.copyBytes(inputStream, outputStream, 4096, true);
    System.out.println("success");
    inputStream.close();
    outputStream.close();
  }
}
```

HDFS 写程序运行结果如图 1-10 所示。

```
root@slave2:~# hadoop jar WriteFile.jar WriteFile
success
root@slave2:~# hadoop fs -ls /user
Found 1 items
-rw-r--r--   1 root supergroup          25 2019-09-25 03:11 /user/demo.txt
```

图 1-10　写程序运行结果图

HDFS 读程序主要代码：

```
public class ReadFile {
  public static void main(String[] args) throws IOException{
    Configuration conf = new Configuration();
    FileSystem fileSystem = FileSystem.get(conf);
    String remote = "/user/demo.txt";
    FSDataInputStream inputStream = fileSystem.open(new Path(dfs));
    IOUtils.copyBytes(inputStream, System.out, 4096, true);
    inputStream.close();
  }
}
```

HDFS 读程序运行结果如图 1-11 所示。

```
root@slave2:~# hadoop jar ReadFile.jar ReadFile
Hello, HDFS!
I'm coming!
```

图 1-11　读程序运行结果图

1.2 HBase的安装

1. 实验目的

学会安装列族数据库 HBase(Hadoop DataBase)的单机版。

2. 实验要求

① Java 8。

② Hadoop 2.2.0。

③ HBase 1.2.3。

④ Windows、Linux、MacOSX 或其他类似于 Unix 的操作系统。

3. 预备知识

HBase[2]是一个可靠性高、性能高、面向列、可伸缩、实时读写的分布式数据库。HBase 不同于一般的关系数据库,它是一个适用于非结构化数据存储的数据库,且 HBase 是基于列的而不是基于行的模式。HBase 利用 HDFS 作为其文件存储系统,利用 Hadoop MapReduce 来处理 HBase 中的海量数据,利用 Zookeeper 作为其分布式协同服务。

HBase 有 3 种安装模式:本地模式、伪分布模式和全分布模式。本地模式是 HBase 的默认模式,该模式是直接在本地文件系统中存放数据而不是基于 HDFS,也不需要使用 Zookeeper。在本地文件系统中运行 HBase 一般不能确保数据的持久性,要想保证数据都被保存下来,需要在 HDFS 上运行 HBase。在本地文件系统中运行 HBase 可以快速学习和熟悉 HBase 系统的运作,因而该模式通常用于 HBase 程序的测试。HBase 的伪分布模式是在一台机器上模拟 HBase 的全分布环境,需要使用 HDFS 来保存数据,同时需要运行 Zookeeper。HBase 的全分布模式通常应用于生产环境中,需要使用多台机器。本实验在本地模式下进行。

4. 实验内容

① 下载 jdk,并解压。

```
tar -axvf jdk-8u221-linux-i586.tar.gz
```

② 下载 HBase,并解压。

```
tar -axvf hbase-2.1.6-bin.tar.gz
```

③ 修改 hbase-2.1.6\conf\hbase-env.sh 文件。

```
export JAVA_HOME = /usr/jdk1.8.0_221
```

④ 通过控制台启动 HBase 程序和关闭 HBase 程序,HBase 的启动如图 1-12 所示。

```
/bin/start-hbase.cmd
/bin/stop-hbase.cmd
```

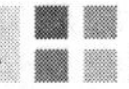

```
root@ubuntu:/usr# hbase-2.1.6/bin/start-hbase.sh
running master, logging to /usr/hbase-2.1.6/bin/../logs/hbase-root-master-ubuntu.out
```

图 1-12　HBase 的启动

⑤ 测试：在浏览器中输入 http://localhost:16010，得到图 1-13 所示的界面。

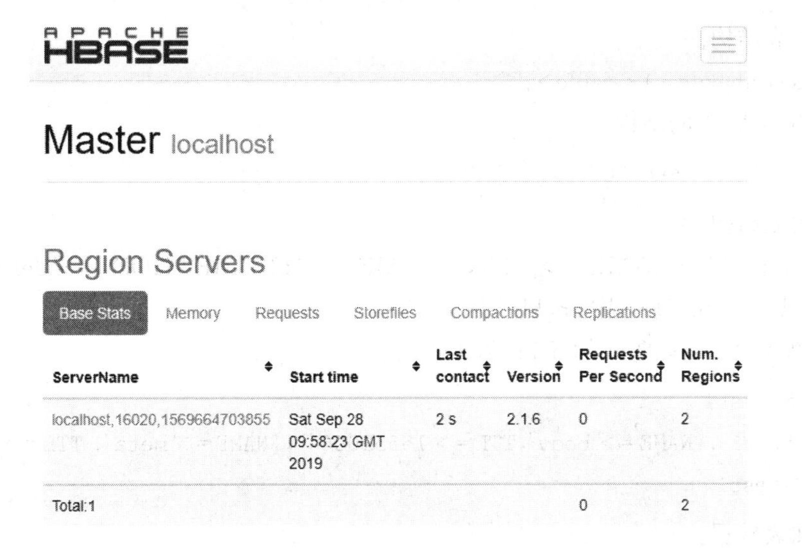

图 1-13　HBase 的监控网页

5. 实验作业

找到 HBase、Hadoop、Java 的安装包，在计算机上安装 HBase。

1.3　HBase 的 Shell 连接与数据操作

1. 实验目的

掌握使用 HBase Shell 连接分布式数据库 HBase 进行建表及增删改查操作。

2. 实验要求

① Linux、MacOSX 或其他类似于 Unix 的操作系统。

② 安装了分布式数据库 HBase 的计算机。

3. 预备知识

HBase 是一个分布式的、面向列的开源数据库，在 Hadoop 之上提供类似于 BigTable 的存储能力，适用于非结构化数据的存储。

4. 实验内容[3]

① 在实验之前需先安装、配置 HBase 及相应的依赖环境，建议使用单节点独立 HBase 模式完成实验，具体安装配置可参考扩展资料。

② 启动 HBase 服务。

语法：bin/start-hbase. sh。

③ 使用 HBase Shell 连接 HBase。

语法:hbase shell。

④ 创建表。

语法:create <table>, {NAME => <family>, VERSIONS => <VERSIONS>}。

实例:创建表 t1,包含两个 family name:f1,f2,且版本数均为 2。

```
create 't1',{NAME => 'f1', VERSIONS => 2},{NAME => 'f2', VERSIONS => 2}
```

⑤ 查看表结构。

语法:describe(desc) <table>。

实例:查看表 t1 的结构。

```
describe 't1' / desc 't1'
```

⑥ 修改表结构。

语法:alter 't1', {NAME => 'f1'}, {NAME => 'f2', METHOD => 'delete'}。修改表结构时必须先"disable"表,再"enable"表。

实例:修改表 test1 的 cf 的 TTL 为 180 天。

```
disable 'test1'
alter 'test1',{NAME => 'body',TTL => '15552000'},{NAME => 'meta', TTL => '15552000'}
enable 'test1'
```

⑦ 添加表数据。

语法:put <table>,<rowkey>,<family:column>,<value>,<timestamp>。

实例:给表 t1 添加一行记录,rowkey 为 rowkey001,family name 为 f1,column name 为 col1,value 为 value01,timestamp 为系统默认。

```
put 't1','rowkey001','f1:col1','value01'
```

⑧ 扫描表数据。

语法:scan <table>, {COLUMNS => [<family:column>, ...], LIMIT => num}。

实例:扫描表 t1 的前 5 条数据。

```
scan 't1',{LIMIT => 5}
```

⑨ 查询单行数据。

语法:get <table>,<rowkey>,[<family:column>, ...]。

实例:查询表 t1 的 rowkey001 中 f1 下的 col1 的值。

```
get 't1','rowkey001', 'f1:col1'
```

或

```
get 't1','rowkey001', {COLUMN => 'f1:col1'}
```

⑩ 删除数据。

a. 删除行中某一个列值。

语法:delete <table>, <rowkey>,<family:column>,<timestamp>。必须指定列名,删除某个列值。

实例:删除表 t1 的 rowkey001 中 f1:col1 的数据。

```
delete 't1','rowkey001','f1:col1'
```

b. 删除某一行数据。

语法：deleteall <table>, <rowkey>, <family:column> , <timestamp>。可以不指定列名，删除整行数据。

实例：删除表 t1 的 rowkey001 的数据。

```
deleteall 't1','rowkey001'
```

c. 清空表数据。

语法：truncate <table>。

实例：删除表 t1 的所有数据。

```
truncate 't1'
```

⑪ 删除表。

语法：第一步，disable <table>；第二步，drop <table>。

实例：删除表 t1。

```
disable 't1'
drop 't1'
```

⑫ 退出 HBase Shell。

语法：quit。

5. 实验作业

① 参考扩展资料，了解 HBase 架构，完成 HBase 的配置和安装。

② 利用 HBase Shell 连接 Hbase，完成表的创建、增删改查等操作。

6. 扩展资料

① 参考链接完成相应 HBase 环境的安装配置和了解。下载地址：http://hbase.apache.org/。

② 使用 HBase Shell 连接 HBase，如图 1-14 所示。

```
[root@slave2 usr]# hbase shell

HBase Shell
Use "help" to get list of supported commands.
Use "exit" to quit this interactive shell.
Version 2.0.0.3.0.1.0-187, re9fcf450949102de5069b257a6dee469b8f5aab3, Wed Sep 19
Took 0.0021 seconds
hbase(main):001:0>
```

图 1-14 HBase Shell 启动界面

③ 创建表，如图 1-15 所示。

```
hbase(main):001:0> create 't1',{NAME=>'f1', VERSIONS=>2},{NAME=>'f2', VERSIONS=>
2}
Created table t1
Took 2.1210 seconds
=> Hbase::Table - t1
```

图 1-15 HBase 创建表

④ 列出表信息，如图 1-16 所示。

```
hbase(main):002:0> describe 't1'
Table t1 is ENABLED
t1
COLUMN FAMILIES DESCRIPTION
{NAME => 'f1', VERSIONS => '2', EVICT_BLOCKS_ON_CLOSE => 'false', NEW_VERSION_BE
HAVIOR => 'false', KEEP_DELETED_CELLS => 'FALSE', CACHE_DATA_ON_WRITE => 'false'
, DATA_BLOCK_ENCODING => 'NONE', TTL => 'FOREVER', MIN_VERSIONS => '0', REPLICAT
ION_SCOPE => '0', BLOOMFILTER => 'ROW', CACHE_INDEX_ON_WRITE => 'false', IN_MEMO
RY => 'false', CACHE_BLOOMS_ON_WRITE => 'false', PREFETCH_BLOCKS_ON_OPEN => 'fal
se', COMPRESSION => 'NONE', BLOCKCACHE => 'true', BLOCKSIZE => '65536'}
{NAME => 'f2', VERSIONS => '2', EVICT_BLOCKS_ON_CLOSE => 'false', NEW_VERSION_BE
HAVIOR => 'false', KEEP_DELETED_CELLS => 'FALSE', CACHE_DATA_ON_WRITE => 'false'
, DATA_BLOCK_ENCODING => 'NONE', TTL => 'FOREVER', MIN_VERSIONS => '0', REPLICAT
ION_SCOPE => '0', BLOOMFILTER => 'ROW', CACHE_INDEX_ON_WRITE => 'false', IN_MEMO
RY => 'false', CACHE_BLOOMS_ON_WRITE => 'false', PREFETCH_BLOCKS_ON_OPEN => 'fal
se', COMPRESSION => 'NONE', BLOCKCACHE => 'true', BLOCKSIZE => '65536'}
2 row(s)
Took 0.2369 seconds
```

图 1-16　HBase 列出表信息

⑤ 修改表结构，如图 1-17 所示。

```
hbase(main):005:0> disable 't1'
Took 1.4423 seconds
hbase(main):006:0> alter 't1',{NAME=>'body',TTL=>'15552000'},{NAME=>'meta', TTL=
>'15552000'}
Updating all regions with the new schema...
All regions updated.
Done.
Took 1.7758 seconds
hbase(main):007:0>  enable 't1'
NoMethodError: undefined method ` enable' for main:Object

hbase(main):008:0> enable 't1'
Took 1.5019 seconds
hbase(main):009:0>
```

图 1-17　HBase 修改表结构

⑥ 放置表数据，如图 1-18 所示。

```
hbase(main):009:0> put 't1','rowkey001','f1:col1','value01'
Took 0.2370 seconds
```

图 1-18　HBase 添加数据

⑦ 扫描表数据，如图 1-19 所示。

```
hbase(main):010:0> scan 't1',{LIMIT=>5}
ROW                    COLUMN+CELL
 rowkey001             column=f1:col1, timestamp=1569652707916, value=value01
1 row(s)
Took 0.1220 seconds
```

图 1-19　HBase 扫描表数据

⑧ 查询单行数据，如图 1-20 所示。

```
hbase(main):011:0> get 't1','rowkey001', {COLUMN=>'f1:col1'}
COLUMN                   CELL
 f1:col1                 timestamp=1569652707916, value=value01
1 row(s)
Took 0.0406 seconds
```

图 1-20　HBase 查询单行数据

⑨ 删除数据行，如图 1-21 和图 1-22 所示。

```
hbase(main):012:0* delete 't1','rowkey001','f1:col1'
Took 0.0503 seconds
```

图 1-21　HBase 删除行中的某一个列值

```
hbase(main):013:0> deleteall 't1','rowkey001'
Took 0.0072 seconds
```

图 1-22　HBase 删除某一行数据

⑩ 清空表数据，如图 1-23 所示。

```
hbase(main):015:0> truncate 't1'
Truncating 't1' table (it may take a while):
Disabling table...
Truncating table...
Took 3.8623 seconds
```

图 1-23　HBase 清空表数据

⑪ 删除表，如图 1-24 所示。

```
hbase(main):016:0> disable "t1"
Took 1.4684 seconds
hbase(main):017:0> drop "t1"
Took 0.8258 seconds
hbase(main):018:0>
```

图 1-24　HBase 删除表

7. 参考答案

实验作业的答案见本节扩展资料部分。

1.4　Redis 数据库的安装和使用

1. 实验目的

拓展阅读

Redis 介绍

① 学会安装键值数据库 Redis。

② 掌握使用 Redis-cli 对键值数据库 Redis 进行增删改查操作。

2. 实验要求

① 安装了 Redis 数据库的计算机。

② Windows、Linux、MacOSX 或其他类似于 Unix 的操作系统。

3. 预备知识

Redis是完全开源和免费的，遵守BSD协议，是一个高性能的键值数据库。Redis支持数据的持久化，可以将内存中的数据保存在磁盘中，重启后可以再次加载使用。Redis不仅支持简单的key-value类型的数据，还提供list、set、zset、hash等数据结构的存储。Redis支持数据的备份，即master-slave模式的数据备份。

4. 实验内容[4]

① 下载Redis安装包并安装。

本书选用的是redis-5.0.5。下载、安装命令如下所示。

```
wget http://download.redis.io/releases/redis-5.0.5.tar.gz
tar xzf redis-5.0.5.tar.gz
cd redis-5.0.5
make
```

② Redis的启动与关闭。

语法：

src/redis-server(服务器启动)；

src/redis-cli(客户端启动)；

shutdown(关闭客户端)。

③ Redis连接命令。

语法：

auth password(验证密码是否正确)；

ping(查看服务器是否运行)；

quit(关闭当前连接)；

select index(切换到指定服务器)。

④ Redis中键的常用命令。

语法：

set key value(设置键值对)；

del key(删除键)；

exists key(检查键是否存在)；

keys pattern(查找所有符合给定模式的键)；

rname key newkey(修改键的名称)；

renamex key newkey(仅当newkey不存在时，将key改名为newkey)；

type key(返回key所存储的值的类型)。

⑤ Redis中字符串的常用命令。

在Redis中，字符串可以存储字节串(byte string)、整数和浮点数三种类型的值。用户可以通过给定一个任意的数值，对存储整数或浮点数的字符串执行自增(increment)或者自减(decrement)操作。在需要的时候，可以将整数转换为浮点数。

语法：

getrange key start end(返回字符串的子字符串)；

getset key value(将给定key的值设置为value，并返回key的旧值)；

strlen key(返回key中存储的字符串值的长度)；

incr key(将key中存储的数字值增一)；

incrby key increment（将 key 中存储的值加上给定的增量值）；
incrbyfloat key increment（将 key 中存储的值加上给定的浮点增量值）；
decr key（将 key 中存储的数字值减一）；
decrby key increment（将 key 中存储的值减去给定的增量值）；
append key value（在字符串末尾增加值）。
实例：

```
set mynum "2"
get mynum
incr mynum
incrby mynum"3"
decr mynum
decrby mynum"4"
```

⑥ Redis 中列表的常用命令。
语法：
lpush key value（在列表的左边插入元素）；
rpush key value（在列表的右边插入元素）；
lrange key value start end（列出列表中[start,end]的元素）。
实例：

```
lpush mylist "1"  //新建一个列表 mylist,并在列表头部插入元素"1"
rpush mylist "2"  //在 mylist 右侧插入元素"2"
lpush mylist "0"  //在 mylist 左侧插入元素"0"
lrange mylist 0 1 //列出 mylist 中从编号 0 到编号 1 的元素
lrange mylist 0 -1 //列出 mylist 中的所有元素
```

⑦ Redis 中集合的常用命令。
语法：
sadd key value（在集合中添加元素）；
smembers key（列出集合中的所有元素）；
sismember key value（判断元素是否在集合中，存在返回 1，不存在返回 0）；
sunion key1 key2（将两个集合合并）。
实例：

```
sadd myset "one"          //在集合 myset 中加入一个新元素"one"
sadd myset "two"          //在集合 myset 中加入一个新元素"two"
smembers myset            //列出集合 myset 中的所有元素
sismember myset "one"     //判断元素"one"是否在集合 myset 中,返回 1 表示存在
sismember myset "three"   //判断元素"three"是否在集合 myset 中,返回 0 表示不存在
sadd yourset "1"
sadd yourset "2"          //新建一个集合 yourset
sunion myset yourset      //将两个集合合并
```

⑧ Redis 中有序集合的常用命令。

语法：

zadd key index value(在有序集合中添加元素和索引)；

zrange ley start end(列出集合中[start,end]的值)。

实例：

```
zadd myzset 1 baidu.com      //在集合 myzset 中新增一个元素 baidu.com,赋予它的序号是 1
zadd myzset 3 360.com        //在集合 myzset 中新增一个元素 360.com,赋予它的序号是 3
zadd myzset 2 google.com     //在集合 myzset 中新增一个元素 google.com,赋予它的序号是 2
zrange myzset 0 -1 with scores //列出 myzset 的所有元素,同时列出其序号,可以看出
                                 myzset 是有序的
zrange myzset 0 -1           //只列出 myzset 的元素
```

⑨ Redis 中哈希的常用命令。

语法：

HMSET key hash1 hash2...(在 hash 中添加元素)；

HGETALL key(列出哈希的元素)；

HSET key hash(修改哈希的值)。

实例：

```
HMSET user:001 username:antirez password:1234 //建立哈希,并赋值
HGETALL user:001                              //列出哈希的内容
HSET user:001 password:0000                   //更改哈希中的某一个值
HGETALL user:001                              //再次列出哈希的内容
```

⑩ Redis 中事务的常用命令。

语法：

MULTI(事务开始)；

INCR command(添加命令)；

EXEC(执行事务)。

实例：

```
MULTI              //标记事务开始
INCR user_id       //多条命令按顺序入队
INCR user_id
INCR user_id
EXEC               //执行
```

注意：在上例的执行中，可以看到 QUEUED 字样，在用 MULTI 组装事务时，每一个命令都会进入内存队列中缓存起来，如果出现 QUEUED 则表示这个命令已成功插入缓存队列，在将来执行 EXEC 时，这些命令会被组装成一个事务来执行。

5. 实验作业

参考扩展资料，完成 Redis 键值数据库的相关操作。

6. 扩展资料

① 启动 Redis 服务，如图 1-25 所示。

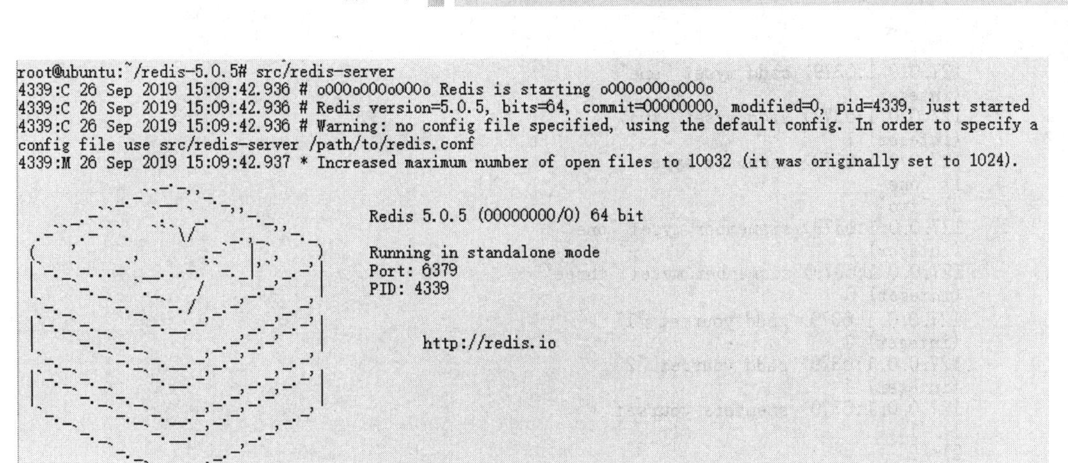

```
root@ubuntu:~/redis-5.0.5# src/redis-server
4339:C 26 Sep 2019 15:09:42.936 # oO0OoO0OoO0Oo Redis is starting oO0OoO0OoO0Oo
4339:C 26 Sep 2019 15:09:42.936 # Redis version=5.0.5, bits=64, commit=00000000, modified=0, pid=4339, just started
4339:C 26 Sep 2019 15:09:42.936 # Warning: no config file specified, using the default config. In order to specify a config file use src/redis-server /path/to/redis.conf
4339:M 26 Sep 2019 15:09:42.937 * Increased maximum number of open files to 10032 (it was originally set to 1024).

Redis 5.0.5 (00000000/0) 64 bit

Running in standalone mode
Port: 6379
PID: 4339

http://redis.io
```

图 1-25　Redis 的启动页面

② 使用 Hbase Shell 连接 HBase。

```
redis-cli.exe -h 127.0.0.1 -p 6379
```

③ Redis 字符串的操作如图 1-26 所示。

```
127.0.0.1:6379> set mynum "2"
OK
127.0.0.1:6379> get mynum
"2"
127.0.0.1:6379> incr mynum
(integer) 3
127.0.0.1:6379> incrby mynum "3"
(integer) 6
127.0.0.1:6379> decr mynum
(integer) 5
127.0.0.1:6379> decrby mynum "4"
(integer) 1
```

图 1-26　Redis 字符串的操作

④ Redis 列表的操作如图 1-27 所示。

```
127.0.0.1:6379> lpush mylist "1"
(integer) 1
127.0.0.1:6379> lpush mylist "2"
(integer) 2
127.0.0.1:6379> lpush mylist "0"
(integer) 3
127.0.0.1:6379> lrange mylist 0 1
1) "0"
2) "2"
127.0.0.1:6379> lrange mylist 0 -1
1) "0"
2) "2"
3) "1"
```

图 1-27　Redis 列表的操作

⑤ Redis 集合的操作如图 1-28 所示。

```
127.0.0.1:6379> sadd myset "one"
(integer) 1
127.0.0.1:6379> sadd myset "two"
(integer) 1
127.0.0.1:6379> smembers myset
1) "one"
2) "two"
127.0.0.1:6379> sismember myset "one"
(integer) 1
127.0.0.1:6379> sismember myset "three"
(integer) 0
127.0.0.1:6379> sadd yourset "1"
(integer) 1
127.0.0.1:6379> sadd yourset "2"
(integer) 1
127.0.0.1:6379> smembers yourset
1) "1"
2) "2"
127.0.0.1:6379> sunion myset yourset
1) "1"
2) "one"
3) "2"
4) "two"
```

图 1-28　Redis 集合的操作

⑥ Redis 有序集合的操作如图 1-29 所示。

```
127.0.0.1:6379> zadd myzset 1 baidu.com
(integer) 1
127.0.0.1:6379> zadd myzset 3 360.com
(integer) 1
127.0.0.1:6379> zadd myzset 2 google.com
(integer) 1
127.0.0.1:6379> zrange myzset 0 -1
1) "baidu.com"
2) "google.com"
3) "360.com"
127.0.0.1:6379> zrange myzset 0 -1 withscores
1) "baidu.com"
2) "1"
3) "google.com"
4) "2"
5) "360.com"
6) "3"
127.0.0.1:6379>
```

图 1-29　Redis 有序集合的操作

⑦ Redis 哈希的操作如图 1-30 所示。

```
127.0.0.1:6379> hset user:001 username:bigdata password:1234
(integer) 1
127.0.0.1:6379> hgetall user:001
1) "username:bigdata"
2) "password:1234"
127.0.0.1:6379> hset user:001 password 0000
(integer) 1
127.0.0.1:6379> hgetall user:001
1) "username:bigdata"
2) "password:1234"
3) "password"
4) "0000"
```

图 1-30　Redis 哈希的操作

⑧ Redis 事务处理如图 1-31 所示。

```
127.0.0.1:6379> multi
OK
127.0.0.1:6379> incr user_id
QUEUED
127.0.0.1:6379> incr user_id
QUEUED
127.0.0.1:6379> incr user_id
QUEUED
127.0.0.1:6379> ping
QUEUED
127.0.0.1:6379> exec
1) (integer) 1
2) (integer) 2
3) (integer) 3
4) PONG
```

图 1-31 Redis 事务处理

1.5 MongoDB 数据库的安装和使用

1. 实验目的

① 学会安装文档数据库 MongoDB。

② 学会使用 MongoDB 运行环境，对文档数据库 MongoDB 进行增删改查操作。

2. 实验要求

① 安装了文档数据库 MongoDB 的计算机。

② Windows、Linux、MacOSX 或其他类似于 Unix 的操作系统。

3. 预备知识

MongoDB 是由 C++语言编写的一个面向文档存储的数据库，操作起来比较简单。MongoDB 旨在为 Web 应用提供可扩展的高性能数据存储解决方案，MongoDB 也是最适合前端学习者学习的数据库。

拓展阅读

MongoDB 介绍

4. 实验内容[5]

① 下载 MongoDB 安装包。

下载地址：https://www.mongodb.com/download-center#community。本实验选择的是 mongodb-linux-x86_64-ubuntu1604-4.2.0.tgz，下载后解压即可。

② MongoDB 服务的启动和退出。

语法：

./bin/mongod(服务器启动)；

./bin/mongo(客户端启动)；

exit(退出客户端)。

③ 批量导入数据。

语法：mongoimport --db database_name --collection collection --file file_name。其中，“mongoimport”为 MongoDB 的导入数据库命令，“--db”指定导入的数据库，如果数据库不存在则会新建，“--collection”指定导入的集合，如果集合不存在则会新建，“--file”指定导入的文件目录。

实例：

```
mongoimport --db student --collection score --file ~/data/test.txt
```

用例 test. txt：

```
{"name":"zhangsan","age":16,"sex":"man","transcript":{"yuwen":90,"shuxue":
                                                    100,"yingyu":90}}
{"name":"lisi","age":16,"sex":"woman","transcript":{"yuwen":89,"shuxue":
                                                    100,"yingyu":90}}
{"name":"wangwu","age":16,"sex":"man","transcript":{"yuwen":85,"shuxue":88,
                                                    "yingyu":79}}
{"name":"zhaoliu","age":16,"sex":"woman","transcript":{"yuwen":90,"shuxue":
                                                    89,"yingyu":77}}
{"name":"yanqi","age":16,"sex":"man","transcript":{"yuwen":78,"shuxue":100,
                                                    "yingyu":90}}
{"name":"shenba","age":16,"sex":"woman","transcript":{"yuwen":78,"shuxue":
                                                    89,"yingyu":89}}
```

④ 通过 mongo 命令使用数据库。

语法：mongo。

注意：后面的所有命令均在该环境下执行。

⑤ 列出所有数据库。

语法：show dbs。

⑥ 使用数据库。

语法：use database_name（database_name 若是一个存在的数据库，则使用这个数据库；若是一个不存在的数据库，则新建一个数据库）。

实例：

```
use student
```

⑦ 插入数据库。

语法：db. userList. insert(data)（向数据库中插入数据，userList 指集合，集合中存储着很多 JSON。需要注意的是 userList 第一次使用时，集合会自动创建，否则会在原来的基础上追加数据）；db. userList. find()（查询数据库中的数据）。

实例：

```
db.userList.insert({name:"zhangsan"})
```

⑧ 删除数据库。

语法：db. dropDatabase()（删除当前使用的数据库）。

⑨ 查找数据。

语法：db. collection. find()（查找该数据库中集合的数据）。

实例：

```
db.score.find()  //find 可以查找数据库 student 中集合为 score 的所有文档
db.score.find({sex:"man"})     //精确查找:查找所有性别为男的文档
db.score.find({sex:"man","transcript.yuwen":78})//多个条件:查找所有性别为男
                                                且语文分数为 78 分的文档
db.score.find({age:{$gt:50}})//大于条件:查找所有年龄大于 50 的文档($gt:大
                        于,$lt:小于,$gte:大于或等于,$lte:小于或等于)
db.score.find({name:/zhang/}) //模糊查找:查找名字带有"zhang"的文档
db.score.find({$or:[{age:12},{age:18}]})  //或:查找年龄为 12 或 18 的文档
db.score.find().sort({age:1}) //查找结果排序:按照年龄的正序排列
db.score.find().limit(2)     //通过 limit 取指定的条数
db.score.find().skip(2)      //跳过前两条记录
```

⑩ 删除数据。

语法:db.score.remove()。

实例：

```
db.score.remove({sex:"woman"})                  //删除所有女生
db.score.remove({sex:"man"},{justOne:true})     //删除一条男生信息
db.score.remove({})                             //删除集合中的所有文档
```

⑪ 显示当前数据库中的所有集合。

语法:show collections。

5. 实验作业

参考扩展资料,完成 MongoDB 文档数据库的相关操作。

6. 扩展资料

① MongoDB 服务的启动和退出,如图 1-32 所示。

```
2019-09-28T11:47:11.848+0800 I CONTROL  [initandlisten] MongoDB starting : pid=9
08 port=27017 dbpath=D:\data\db\ 64-bit host=DESKTOP-MB1CH9H
2019-09-28T11:47:11.851+0800 I CONTROL  [initandlisten] targetMinOS: Windows 7/W
indows Server 2008 R2
2019-09-28T11:47:11.853+0800 I CONTROL  [initandlisten] db version v3.2.22
2019-09-28T11:47:11.853+0800 I CONTROL  [initandlisten] git version: 105acca0d44
3f9a47c1a5bd608fd7133840a58dd
2019-09-28T11:47:11.853+0800 I CONTROL  [initandlisten] OpenSSL version: OpenSSL
 1.0.2o-fips  27 Mar 2018
2019-09-28T11:47:11.853+0800 I CONTROL  [initandlisten] allocator: tcmalloc
2019-09-28T11:47:11.853+0800 I CONTROL  [initandlisten] modules: none
2019-09-28T11:47:11.853+0800 I CONTROL  [initandlisten] build environment:
2019-09-28T11:47:11.853+0800 I CONTROL  [initandlisten]     distmod: 2008plus-ss
l
2019-09-28T11:47:11.853+0800 I CONTROL  [initandlisten]     distarch: x86_64
2019-09-28T11:47:11.853+0800 I CONTROL  [initandlisten]     target_arch: x86_64
2019-09-28T11:47:11.853+0800 I CONTROL  [initandlisten] options: {}
2019-09-28T11:47:11.855+0800 I STORAGE  [initandlisten] exception in initAndList
en: 29 Data directory D:\data\db\ not found., terminating
2019-09-28T11:47:11.855+0800 I CONTROL  [initandlisten] dbexit:  rc: 100
```

图 1-32 MongoDB 启动页面

② 批量导入数据,如图 1-33 所示。

```
2019-09-28T10:39:14.468+0800    connected to: localhost
2019-09-28T10:39:14.519+0800    imported 6 documents
```

图 1-33　MongoDB 批量导入数据

③ 通过 mongo 命令使用数据库,如图 1-34 所示。

```
MongoDB shell version: 3.2.22
connecting to: test
Server has startup warnings:
2019-09-24T22:16:15.948+0800 I CONTROL  [initandlisten]
2019-09-24T22:16:15.948+0800 I CONTROL  [initandlisten] ** WARNING: Access contr
ol is not enabled for the database.
2019-09-24T22:16:15.948+0800 I CONTROL  [initandlisten] **          Read and wri
te access to data and configuration is unrestricted.
2019-09-24T22:16:15.948+0800 I CONTROL  [initandlisten]
>
```

图 1-34　MongoDB shell 启动界面

④ 列出所有数据库,如图 1-35 所示。

```
> show dbs
admin    0.000GB
config   0.000GB
local    0.000GB
student  0.000GB
>
```

图 1-35　MongoDB 列出所有数据库

⑤ 使用数据库,如图 1-36 所示。

```
> use student
switched to db student
>
```

图 1-36　MongoDB shell 使用数据库

⑥ 插入数据库,如图 1-37 所示。

```
> use class
switched to db class
> db
class
> db.userList.insert({name:"zhangsan"})
WriteResult({ "nInserted" : 1 })
> db.userList.find()
{ "_id" : ObjectId("5d8ec5a3a02279ff9d821121"), "name" : "zhangsan" }
>
```

图 1-37　MongoDB shell 插入数据库

⑦ 删除数据库，如图 1-38 所示。

```
> show dbs
admin     0.000GB
calss     0.000GB
config    0.000GB
local     0.000GB
student   0.000GB
> db.dropDatabase()
{ "dropped" : "calss", "ok" : 1 }
> show dbs
admin     0.000GB
config    0.000GB
local     0.000GB
student   0.000GB
>
```

图 1-38　MongoDB shell 删除数据库

⑧ 查找数据，如图 1-39 所示。

```
> use student
switched to db student
> db.score.find()
{ "_id" : ObjectId("5e018541089d6d7ebbf998ef"), "name" : "lisi", "age" : 16, "sex" :
"woman", "transcript" : { "yuwen" : 89, "shuxue" : 100, "yingyu" : 90 } }
{ "_id" : ObjectId("5e018541089d6d7ebbf998f0"), "name" : "wangwu", "age" : 16, "sex"
: "man", "transcript" : { "yuwen" : 85, "shuxue" : 88, "yingyu" : 79 } }
{ "_id" : ObjectId("5e018541089d6d7ebbf998f1"), "name" : "zhangsan", "age" : 16, "se
x" : "man", "transcript" : { "yuwen" : 90, "shuxue" : 100, "yingyu" : 90 } }
{ "_id" : ObjectId("5e018541089d6d7ebbf998f2"), "name" : "zhaoliu", "age" : 16, "sex
" : "woman", "transcript" : { "yuwen" : 90, "shuxue" : 89, "yingyu" : 77 } }
{ "_id" : ObjectId("5e018541089d6d7ebbf998f3"), "name" : "yanqi", "age" : 16, "sex"
: "man", "transcript" : { "yuwen" : 78, "shuxue" : 100, "yingyu" : 90 } }
{ "_id" : ObjectId("5e018541089d6d7ebbf998f4"), "name" : "shenba", "age" : 16, "sex"
: "woman", "transcript" : { "yuwen" : 78, "shuxue" : 89, "yingyu" : 89 } }
```

图 1-39　MongoDB shell 查找数据

⑨ 删除数据，如图 1-40 所示。

```
> use student
switched to db student
> db.score.remove({sex:"woman"})
WriteResult({ "nRemoved" : 3 })
> db.score.find()
{ "_id" : ObjectId("5e018541089d6d7ebbf998f0"), "name" : "wangwu", "age" : 16, "sex"
: "man", "transcript" : { "yuwen" : 85, "shuxue" : 88, "yingyu" : 79 } }
{ "_id" : ObjectId("5e018541089d6d7ebbf998f1"), "name" : "zhangsan", "age" : 16, "se
x" : "man", "transcript" : { "yuwen" : 90, "shuxue" : 100, "yingyu" : 90 } }
{ "_id" : ObjectId("5e018541089d6d7ebbf998f3"), "name" : "yanqi", "age" : 16, "sex"
: "man", "transcript" : { "yuwen" : 78, "shuxue" : 100, "yingyu" : 90 } }
>
> db.score.remove({sex:"man"},{justOne:true})
WriteResult({ "nRemoved" : 1 })
> db.score.find()
{ "_id" : ObjectId("5e018541089d6d7ebbf998f1"), "name" : "zhangsan", "age" : 16, "se
x" : "man", "transcript" : { "yuwen" : 90, "shuxue" : 100, "yingyu" : 90 } }
{ "_id" : ObjectId("5e018541089d6d7ebbf998f3"), "name" : "yanqi", "age" : 16, "sex"
: "man", "transcript" : { "yuwen" : 78, "shuxue" : 100, "yingyu" : 90 } }
>
> db.score.remove({})
WriteResult({ "nRemoved" : 2 })
> db.score.find()
>
```

图 1-40　MongoDB shell 删除数据

⑩ 查找集合，如图 1-41 所示。

```
> use student
switched to db student
> show collections
score
> _
```

图 1-41　MongoDB shell 查找集合

1.6　Neo4j 数据库的安装和使用

1. 实验目的

① 学会在计算机上安装图数据库 Neo4j。

② 学会使用 Cypher 语言对图数据库 Neo4j 进行增删改查操作。

2. 实验要求

① 安装了图数据库 Neo4j 的计算机。

② Java 8。

③ Windows、Linux、MacOSX 或其他类似于 Unix 的操作系统。

3. 预备知识

Neo4j 是一个高性能的 NoSQL 图形数据库，它是一个嵌入式的、基于磁盘的、具备完全的事务特性的 Java 持久化引擎，但它是将结构化数据存储在网络(从数学角度叫作图)上而不是表中。Neo4j 也可以看作一个高性能的图引擎，该引擎具有成熟数据库的所有特性。

4. 实验内容[6]

① 在官网下载 Neo4j 安装包。

下载网站：https://neo4j.com/download-center/#releases。或者直接使用命令：

```
curl -O http://dist.neo4j.org/neo4j-community-3.4.5-unix.tar.gz
```

② 解压启动 Neo4j 服务器。

解压：

```
tar -axvf neo4j-community-3.4.5-unix.tar.gz
```

启动：

```
./neo4j-community-3.4.5/bin/neo4j start
```

停止：

```
./neo4j-community-3.4.5/bin/neo4j stop
```

③ 打开 Neo4j 集成的浏览器。在一个运行的服务器实例上访问“http://localhost:7474/”，打开浏览器，显示启动页面，默认的用户是 neo4j，其默认的密码为 neo4j，第一次成功登录 Neo4j 之后，需要重置密码，如图 1-42 所示(注：后面所有的命令均在该浏览器内执行)。

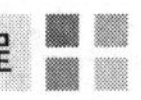

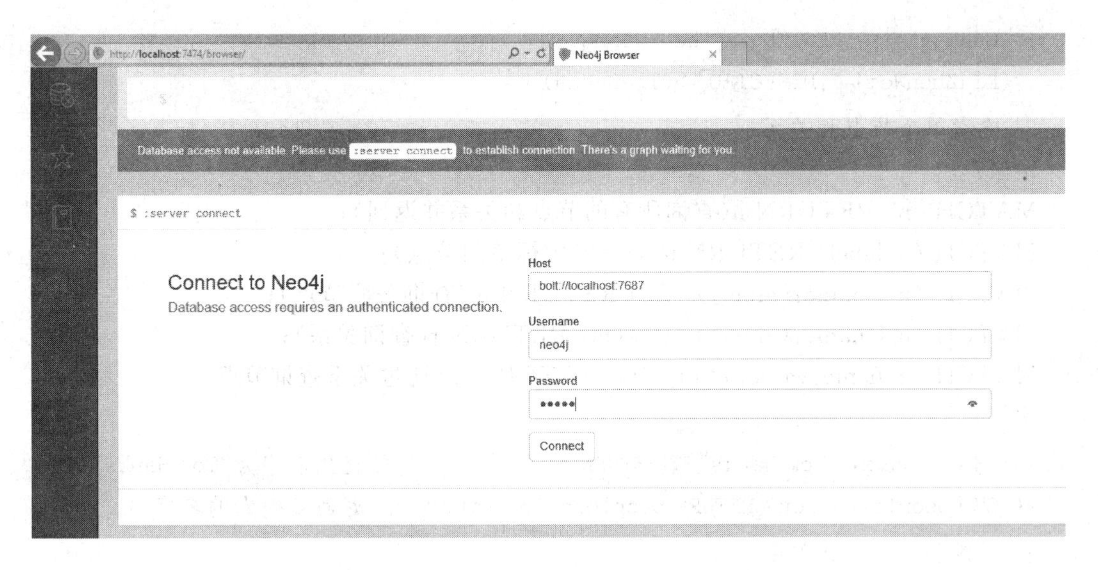

图 1-42 Neo4j 浏览器操作界面

④ 创建节点与关系。[7]

语法：

CREATE (n)(创建一个节点)；

CREATE (n:label)(创建带有标签的节点)；

CREATE (n:label1:label2)(创建带有两个标签的节点,标签可以有多个)；

CREATE (n:label {name1:value1, name2:value2})(创建带有一个标签两个属性的节点,属性可以有多个)；

CREATE (n)-[r:type]->(m)(创建带有一个类型的关系,类型只能有一个,"->"表示关系是从 n 指向 m 的)；

CREATE (n)-[r:type{name1:value1, name2:value2}]->(m)(创建带有两个属性的关系,属性可以有多个)。

实例：

```
CREATE (TheLiuLang:Movie {title:'liulangdiqiu', released:2019, tagline:'liu lang
                          di qiu'})
CREATE (TheDaHua:Movie {title:'dahuaxiyou', released:1994, tagline:'da hua xi you'})
CREATE (WuJing:Person {name:'wujing', born:1974})
CREATE (GuoFan:Person {name:'guofan', born:1980})
CREATE (WuMengDa:Person {name:'WuMengda', born:1953})
CREATE (ZhouXingChi:Person {name:'zhouxingchi', born:1962})
CREATE (LiuZhenWei:Person {name:'liuzhenwei', born:1952})
CREATE
  (WuJing)-[:ACTED_IN {roles:['LiuPeiQiang']}]->(TheLiuLang),
  (WuMengDa)-[:ACTED_IN {roles:['HaiZiAng']}]->(TheLiuLang),
  (WuMengDa)-[:ACTED_IN {roles:['ErDangJia']}]->(TheDaHua),
  (ZhouXingChi)-[:ACTED_IN {roles:['ZhiZongBao']}]->(TheDaHua),
```

```
  (GuoFan)-[:DIRECTED]->(TheLiuLang),
  (LiuZhenWei)-[:DIRECTED]->(TheDaHua)
```

⑤ 检索单个节点和关系。

语法：

MATCH (n) RETURN n(查询所有的节点和关系并返回)；

MATCH (n:label) RETURN n(查询指定标签的节点)；

MATCH (n { name:value })--(m) RETURN m(查询关联节点)；

MATCH (n { name:value })-[r]->(m) RETURN r(查询关系)；

MATCH (n{name:value})<-[r]-(m) RETURN m(通过关系查询节点)。

实例：

```
MATCH (n {name: "Tom Hanks"}) RETURN n              //查找名字为"Tom Hanks"的人
MATCH (people:Person) RETURN people.name LIMIT 4   //返回 4 个人的名字
```

⑥ 寻找多个节点和关系。

实例：

```
MATCH (n:Person {name: "WuMengda"})-[:ACTED_IN]->(m) RETURN n,m //列出吴孟达参演
                                                                 的所有电影
MATCH (n {title: "liulangdiqiu"})<-[:DIRECTED]-(director) RETURN director.name
                                    //寻找指导了流浪地球的导演
MATCH (n:Person {name:"WuMengda"})-[:ACTED_IN]->(m)<-[:ACTED_IN]-(x) RETURN x.name
                                    //寻找和吴孟达拍过戏的演员
```

⑦ 通过图数据库推荐。

情景:为郭凡推荐新的合作演员。一个基本的方法是找到一个相邻的、与他有着良好的联系的人。例如,找到郭凡还没有合作过,但与他合作过的其他演员认识的演员;找一个能把这个演员介绍给他潜在的合作者的人。

实例：

```
MATCH (n:Person {name:"guofan"})-[:DIRECTED]->(m1)<-[:ACTED_IN]-(x),
(x)-[:ACTED_IN]->(m2)<-[:ACTED_IN]-(y)
WHERE NOT (y)-[:ACTED_IN]->(m1)
RETURN y.name AS Recommended  //通过郭凡的合作演员,寻找没有和他合作过的演员
MATCH (n:Person {name:"guofan"})-[:DIRECTED]->(m1)<-[:ACTED_IN]-(x),
(x)-[:ACTED_IN]->(m2)<-[:ACTED_IN]-(y)
RETURN x.name                 //找到一个人把周星驰介绍给郭凡
```

⑧ 删除图数据库。

语法:DELETE。同时删除节点和关系(关系存在,节点不会被删除)。

实例：

```
MATCH (a:Person),(m:Movie) OPTIONAL MATCH (a)-[r1]->(), (m)-[r2]->()
DELETE a,r1,m,r2              //删除所有的电影人物节点和他们之间的关系
```

5. 实验作业

参考扩展资料,完成 Neo4j 图数据库的相关操作。

6. 扩展资料

① 创建电影人物节点与关系，如图 1-43 所示。

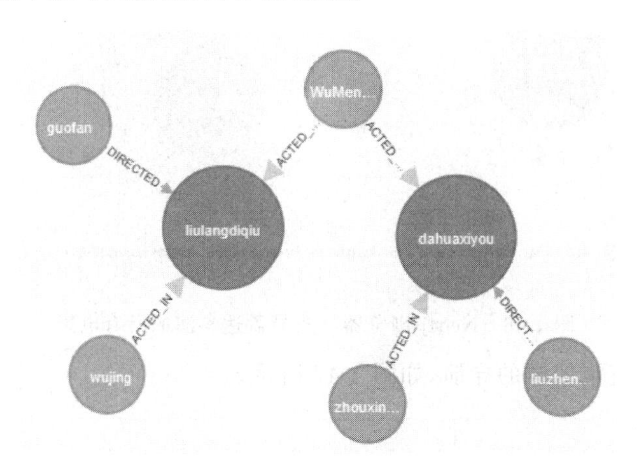

图 1-43 Neo4j 浏览器创建数据界面

② 检索单个节点与关系。例如，寻找周星驰如图 1-44 所示。

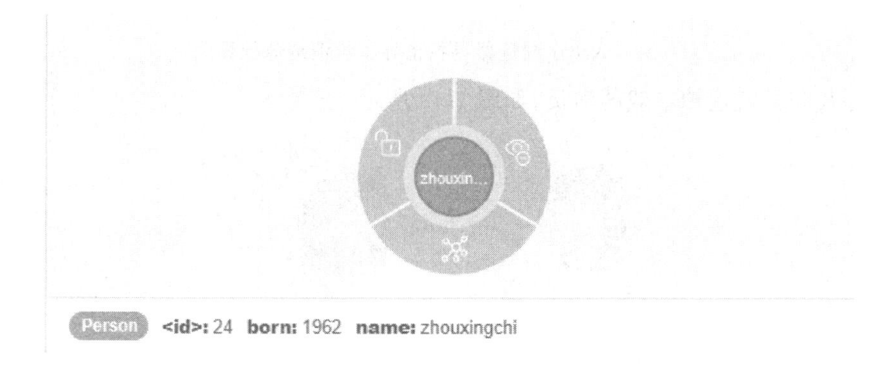

图 1-44 Neo4j 浏览器寻找周星驰结果

再如，寻找 4 个人的名字，如图 1-45 所示。

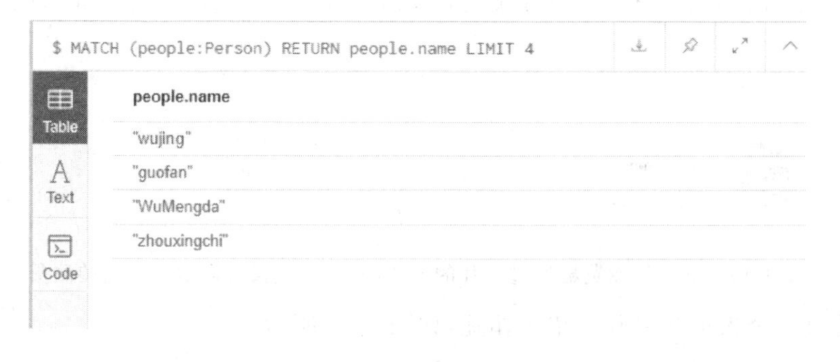

图 1-45 Neo4j 浏览器寻找结果

③ 寻找多个节点和关系。例如，列出吴孟达参演的所有电影，如图 1-46 所示。

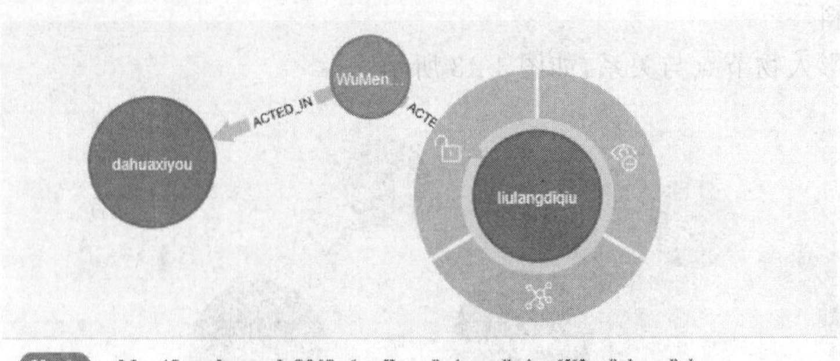

图 1-46　Neo4j 浏览器寻找吴孟达参演的所有电影

再如，寻找指导流浪地球的导演，如图 1-47 所示。

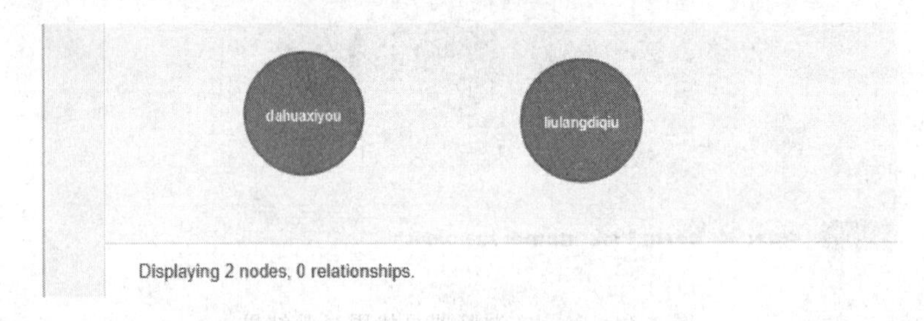

图 1-47　Neo4j 浏览器寻找指导了流浪地球的导演

又如，寻找和吴孟达拍过戏的演员，如图 1-48 所示。

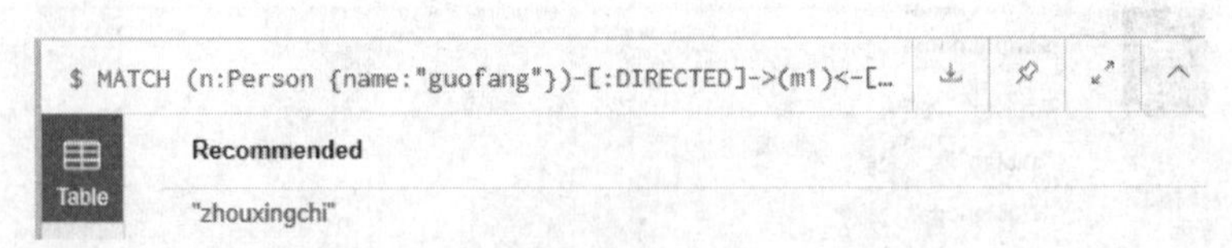

图 1-48　Neo4j 浏览器寻找和吴孟达拍过戏的演员

④ 通过图数据库推荐。例如，通过郭凡的合作演员，寻找没有和他合作过的演员，如图 1-49 所示。

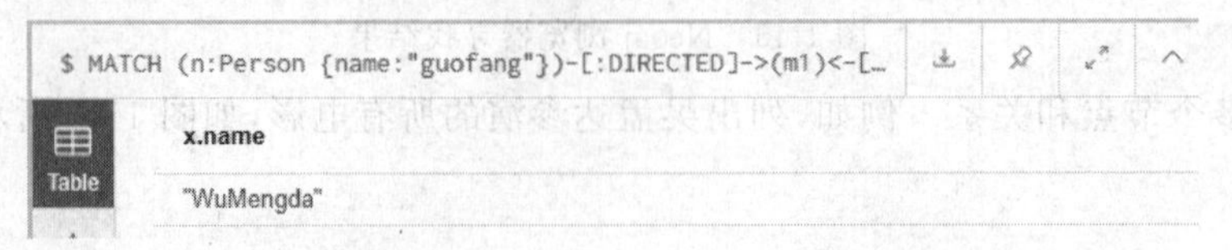

图 1-49　Neo4j 浏览器通过郭凡的合作演员，寻找没有和他合作过的演员

再如，通过一个人把周星驰介绍给郭凡，如图 1-50 所示。

```
$ MATCH (n:Person {name:"guofang"})-[:DIRECTED]->(m1)<-[…
```

x.name
"WuMengda"

图 1-50　Neo4j 浏览器通过一个人把周星驰介绍给郭凡

第2章

大数据处理:MapReduce 处理框架实验教程

拓展阅读

批处理框架

2.1 实验目的

① 通过实验掌握基本的 MapReduce 编程方法。

② 学会用 MapReduce 解决一些常见的数据处理问题,包括数据去重、数据排序和数据自定义格式输出等。

③ 通过操作 MapReduce 的实验,模仿实验内容,深入理解 MapReduce 的过程,熟悉 MapReduce 程序的编程方式。

2.2 实验要求

① Linux 64 位操作系统(CentOS 7 或 RHEL 7,不支持 Windows)。

② JDK 1.8。

③ Hadoop 3.1.2。

④ IDE:Eclipse。

2.3 预备知识

1. MapReduce 简介

MapReduce 是一个经典的用于处理海量数据的分布式批处理计算引擎[8],解决了数据分布式存储、作业调度、容错、机器间通信等复杂问题。

2. MapReduce 运行过程

(1) Map 任务处理

① 读取 HDFS 上的文件,每一行通过 InputSplit 解析成一个<k,v>,每个 InputSplit 都会分配一个 Mapper 任务,每个<k,v>会调用一次 map 函数,如<line0, abcc>,<line1, def>。

拓展阅读

MapReduce 并行处理的基本过程

② 覆盖 map(),接收上述<k,v>,转换为新的<k,v>输出,如<a,1>, <c,1>,<c,1>,<d,1>,<e,1>。

上面的输出会先存放在缓存中,每个 map 都有一个环形内存缓冲区用于存储任务输出(默认大小为 100 MB,io. sort. mb 属性指定),到达阈值 0. 8(io. sort. spill. percent)就溢写到指定的本地目录中。

③ 对②的输出在溢写到磁盘前进行分区(partition),默认是一个分区,分区数量根据 reduce 的数量来取模,如<0,a,1>,<0,b,1>,<0,c,1>,<0,c,1>。

④ 分区后按照<k,v>中的 k 排序以及分组,分组是指将相同 key 的 value 放到一个集合中。排序后:<a,1>,<b,1>,<c,1>,<c,1>;分组后:<a,{1}>,<b,{1}>,<c,{1,1}>。

⑤ (可选)对分组后的数据进行归约(combine)。

(2) Reduce 任务处理

① 多个 Map 任务的输出,按照不同的分区,通过网络复制到不同的 Reduce 节点上进行洗牌(Shuffle)。

② 从 Map 端复制过来的数据首先会写到 Reduce 端的缓存中,缓存占用到达一定的阈值后会写到磁盘中进行分区、合并、排序等过程,如果形成了多个磁盘文件还会进行合并,最后一次合并的结果是作为 Reduce 的输入而不是写入磁盘中。

③ 最后一次合并的结果作为输入传入 Reduce 任务中,当 Reduce 输入文件确定后,整个 Shuffle 操作才算最终结束,之后就是 Reduce 的计算,并把结果存到 HDFS 上。

2.4 实验内容(5 个实验)

1. 实验 1:单词计数

(1) 实验目的

基于 MapReduce 思想,编写单词计数程序。

(2) 实验要求

① 理解 MapReduce 编程思想。

② 编写 MapReduce 版 WordCount 程序。

③ 知悉执行过程。

拓展阅读

优化洗牌(Shuffle)和排序阶段

(3) 实验原理

① MapReduce 运行。

MapReduce 分为 Map 和 Reduce,一个 Map 处理一个数据块,所以每个机器上会有多个 Map,用来处理存储在这个机器上的多个数据块,处理的结果形成 key/value 键值对的形式。Map 处理后的结果由 Reduce 汇总,最后将最终结果进行输出。例如,计算文本中单词出现的个数(WordCount),如图 2-1 所示。

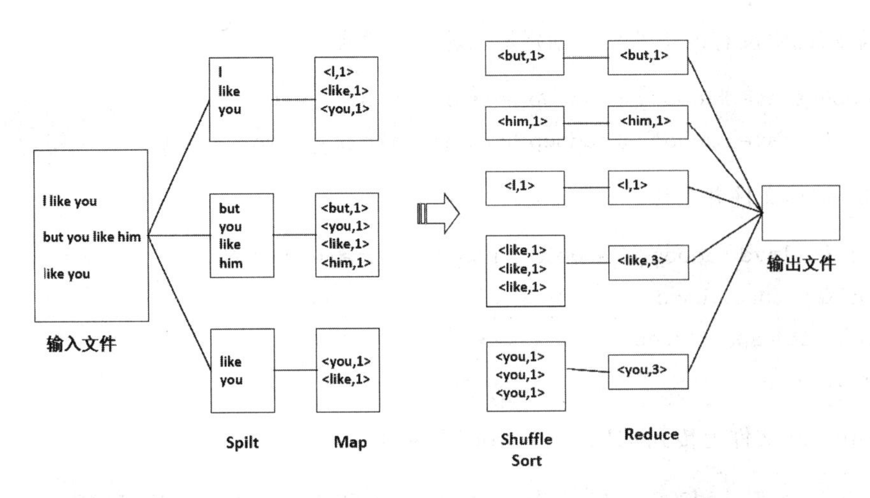

图 2-1　MapReduce 对 WordCount 的处理模式

输入文件分成 3 个数据块，每个 Map 对应一个数据块，对它们进行处理，这里将每个单词出现的次数先置为 1，然后进行洗牌、排序。所谓的洗牌就是将 key 值（这里是单词）相同的放到一起，排序就是按照 key 进行排序，如 5 个键值就是按照“b，h，i，l，y”顺序排好的。从上到下来看，but 排在最上面，you 在最下面。Reduce 就是将 key 值出现的次数进行汇总，把 value 值进行相加，其结果就是这个单词出现的次数，最后将总的结果输出到文件中。

② Java API 解析。[9]

a. InputFormat：用于描述输入数据的格式，常用的为 TextInputFormat，提供如下两个功能。

数据切分：按照某个策略将输入数据切分成若干个 split，以便确定 Map Task 的个数以及对应的 split。

为 Mapper 提供数据：给定某个 split，能将其解析成 key/value 对。

b. OutputFormat：用于描述输出数据的格式，它能够将用户提供的 key/value 对写入特定格式的文件中。

c. Mapper/Reducer：Mapper/Reducer 中封装了应用程序的数据处理逻辑。

d. Writable：Hadoop 自定义的序列化接口，实现该类的接口可以用作 MapReduce 过程中的 value 数据。

e. WritableComparable：在 Writable 的基础上继承了 Comparable 接口，实现该类的接口可以用作 MapReduce 过程中的 key 数据（因为 key 包含了比较排序的操作）。

（4）实验步骤

① 执行命令启动 Hadoop。

```
[hadoop@master ~]$ cd /modules/hadoop/
[hadoop@master hadoop]$ sbin/start-all.sh
```

② 查看 HDFS 上是否存在 WordCount 文件夹。

```
[hadoop@slave ~]$ cd /modules/Hadoop/
[hadoop@slave hadoop]$ bin/Hadoop dfs -ls /
[hadoop@slave hadoop]$
```

没有显示说明没有该文件夹。创建输入输出文件夹：

```
[hadoop@slave hadoop]$ hadoop hdfs dfs -mkdir -p /word
[hadoop@slave hadoop]$ hadoop hdfs dfs -mkdir -p /wordcount
```

③ 编写数据文件并上传。

```
[hadoop@slave hadoop]$ sudo vi /modules/data/word.txt
Hadoop MapReduce hive
Sqoop hadoop spark hbase
Storm hive
```

将 word.txt 文件上传到 HDFS 的/word 目录中：

```
[hadoop@slave hadoop]$ hdfs dfs -put /modules/data/word.txt /word
```

④ 编写 MapReduce 程序。主要编写 Map 和 Reduce 类，Map 过程需要继承 Mapper 类，并重写 map 方法；Reduce 过程需要继承 Reduce 类，并重写 reduce 方法。

```
import java.io.IOException;
import org.apache.hadoop.io.IntWritable;
import org.apache.commons.lang.StringUtils;
import org.apache.hadoop.io.LongWritable;
import org.apache.hadoop.io.Text;
import org.apache.hadoop.mapreduce.Mapper;
import org.apache.hadoop.mapreduce.Reducer;
import org.apache.hadoop.conf.Configuration;
import org.apache.hadoop.fs.Path;
import org.apache.hadoop.mapreduce.lib.input.FileInputFormat;
import org.apache.hadoop.mapreduce.lib.output.FileOutputFormat;
import org.apache.hadoop.mapreduce.Job;
import java.util.StringTokenizer;

public class WordCount{

public class WordCountMapper extends Mapper<LongWritable,Text,Text,IntWritable>{

    /*
    *map 方法是提供给 Map Task 进程来调用的，Map Task 进程每读取一行文本就调用
     一次自定义的 map 方法
    *Map Task 在调用 map 方法时，传递的参数：
    *      一行的起始偏移量 LongWritable 作为 key
    *      一行的文本内容 Text 作为 value
    */
```

```
    @Override
    protected void map(LongWritable key, Text value, Context context) throws
IOException, InterruptedException {
        // 拿到一行文本内容,转换成 String 类型
        String line = value.toString();
        // 将这行文本切分成单词
        String[] words = line.split(" ");

        //输出<单词,1>
        for(String word:words){
            context.write(new Text(word), new IntWritable(1));
        }
    }
}

public class WordCountReducer extends Reducer<Text, IntWritable, Text, IntWritable>{
    @Override
    /*
     * reduce 方法提供给 Reduce Task 进程来调用
     *
     * Reduce Task 会将 Shuffle 阶段分发过来的大量 kv 数据对进行聚合,聚合的机制
       是相同 key 的 kv 对聚合为一组
     * 然后 Reduce Task 对每一组聚合 kv 调用一次我们自定义的 reduce 方法
     * 例如:<hello,1><hello,1><hello,1><tom,1><tom,1><tom,1>
     * hello 组会调用一次 reduce 方法进行处理,tom 组也会调用一次 reduce 方法进行处理
     *  调用时传递的参数:
     *        key:一组 kv 中的 key
     *        values:一组 kv 中所有 value 的迭代器
     */
    protected void reduce(Text key, Iterable<IntWritable> values, Context
context) throws IOException, InterruptedException {
        //定义一个计数器
        int count = 0;
        /通过 value 这个迭代器,遍历这一组 kv 中所有的 value,进行累加
        for(IntWritable value:values){
            count + = value.get();
        }
```

```
            //输出这个单词的统计结果
            context.write(key, new IntWritable(count));
        }
    }

    public static void main(String[] args) throws IOException, ClassNotFoundException,
InterruptedException {
            Configuration conf = new Configuration();
            Job wordCountJob = Job.getInstance(conf);
            //重要:指定本 job 所在的 jar 包
            wordCountJob.setJarByClass(WordCountJobSubmitter.class);

            //设置 wordCountJob 所用的 Mapper 逻辑类为哪个类
            wordCountJob.setMapperClass(WordCountMapper.class);
            //设置 wordCountJob 所用的 Reducer 逻辑类为哪个类
            wordCountJob.setReducerClass(WordCountReducer.class);

            //设置 Map 阶段输出数据类型
            wordCountJob.setMapOutputKeyClass(Text.class);
            wordCountJob.setMapOutputValueClass(IntWritable.class);

            //设置最终输出数据类型
            wordCountJob.setOutputKeyClass(Text.class);
            wordCountJob.setOutputValueClass(IntWritable.class);

            //设置要处理的文本数据所存放的路径
             FileInputFormat.setInputPaths(wordCountJob, "hdfs://192.168.31.70:
                                          9000/word/");
             FileOutputFormat.setOutputPath(wordCountJob, new Path("hdfs://192.
                                          168.31.70:9000/wordcount/"));

            //提交 job 给 Hadoop 集群
            wordCountJob.waitForCompletion(true);
        }
    }
```

⑤ 使用 Eclipse 开发工具将该代码打包。打开命令行窗口,切换到工程项目目录下,执行 mvn 打包命令[10]"mvn clean package"。

```
D:\Eclipse\Project\WordCount>mvn clean package
```

看到“BUILD SUCCESS”即打包成功，在项目的 tar 目录内可以进行查看。

(5) 实验结果

在 HDFS 上查看结果。

将项目打成 jar 包后上传到虚机上，并运行 jar 文件：

```
[hadoop@slave hadoop]$
hadoop jar wordcount.jar com.wan.bigdata.wordcount.WordCount /word
/wordcount/output
[hadoop@slave hadoop]$ hadoop dfs -ls /wordcount/output
-rw-r--r-- 1 hadoop supergroup 0 2019-08-06 15:00 /wordcount/output/_SUCCESS
-rw-r--r-- 1 hadoop supergroup 60 2019-08-06 15:00 wordcount/output/part-r-00000
```

查看文件内容：

```
[hadoop@slave hadoop]$ hadoop dfs -cat / wordcount/output/part-r-00000
hadoop   2
Hbase    1
Sqoop    1
Spark    1
Storm    1
MapReduce   1
Hive   2
```

2. 实验 2：去重

(1) 实验目的

基于 MapReduce 思想编写去重程序，使得原始数据中出现次数超过一次的数据在输出文件中只出现一次。

(2) 实验内容

这里使用从某电商网站的数据文件中截取的一部分数据，文件名为 user，文件中包含用户 ID，商品 ID，联系电话三个字段，数据内容以“\t”分割。内容如下：

```
用户 ID  商品 ID  联系电话
1181     100371   18810134342
1001     101487   15810132322
1001     101450   15810134341
1042     101258   15234342342
1067     102051   18342455332
1056     103179   18564456334
1056     103180   17956456442
1056     103182   17345389993
1054     102310   17888884352
1055     101569   18235464642
1054     110565   18106533552
```

```
1054    102319   18154696322
1076    102317   18234345335
1054    103216   18345353312
1056    102310   18222244456
1064    102312   18810435342
1056    103056   18853453353
1056    103045   18854353535
1056    110073   18853435324
1056    102312   18886754544
1056    103090   18645646463
1056    103084   18853456768
1056    103054   18978567453
1056    110068   18853535353
1076    103091   15353535343
1076    103093   18556467564
1076    103090   16995643452
1076    103056   15145465646
1054    103093   15754653533
1054    103090   13464564773
```

编写 MapReduce 版去重程序，根据商品 ID 进行去重。

(3) 实验原理

“数据去重”就是对数据进行有意义的筛选。数据去重涉及很多任务，如统计数据种类的个数、网站日志等。

数据去重的目标是让原始数据中出现次数超过一次的数据在输出文件中只出现一次。在 MapReduce 流程中，Map 的输出<key,value>经过 Shuffle 过程聚集成<key,value-list>后交给 Reduce。根据 Reduce 的过程特性，会自动根据 key 来计算输入的 value 集合，把数据作为 key 输出给 Reduce，无论这个数据出现多少次，只要在最终结果中输出一次即可。具体来说，Reduce 的输入应该以数据作为 key，对 value 则没有要求（可以设置为空）。当 Reduce 接收到一个<key,value>时就直接将输入的 key 复制到输出的 key 中，并将 value 设置成空值，然后输出<key,value>。

(4) 实验步骤

① 开启 Hadoop。

```
[hadoop@master ~]$ cd /modules/hadoop/
[hadoop@master hadoop]$ sbin/start-all.sh
```

② 在 HDFS 上新建文件目录。

```
[hadoop@slave hadoop]$ hdfs dfs -mkdir -p /mymapreduce2/input
[hadoop@slave hadoop]$ hdfs dfs -mkdir -p /mymapreduce2/output
```

③ 上传数据文件到 HDFS。在 Linux 本地创建/data/buyer_favorite1 文件，并导入

HDFS 的/mymapreduce2/input 的目录中。

```
[hadoop@slave hadoop]$ bin/hadoop dfs -put /data/buyer_favorite1 /mymapreduce2/input
```

④ 用 Java 编写去重程序。

```
import java.io.IOException;

import org.apache.hadoop.conf.Configuration;
import org.apache.hadoop.fs.FileSystem;
import org.apache.hadoop.fs.Path;
import org.apache.hadoop.io.NullWritable;
import org.apache.hadoop.io.Text;
import org.apache.hadoop.mapreduce.Job;
import org.apache.hadoop.mapreduce.Mapper;
import org.apache.hadoop.mapreduce.Reducer;
import org.apache.hadoop.mapreduce.lib.input.FileInputFormat;
import org.apache.hadoop.mapreduce.lib.input.TextInputFormat;
import org.apache.hadoop.mapreduce.lib.output.FileOutputFormat;
import org.apache.hadoop.mapreduce.lib.output.TextOutputFormat;

import Utils.FindHDFSText;

public class QuChong{
    public static class Map extends Mapper<Object , Text , Text , NullWritable>{

    private static Text newKey = new Text();
    //实现 map 函数
    public void map(Object key,Text value,Context context) throws IOException,
InterruptedException{

    String line = value.toString();
    System.out.println("line 是:" + line);
    if(line! = null){
        String arr[] = line.split(" ");
        System.out.println("a[1]是" + arr[1]);
        newKey.set(arr[1]);
        context.write(newKey, NullWritable.get());
        System.out.println("新的截取的值是" + newKey); }
    }
    }
```

```
        //Reduce 将输入中的 key 复制到输出数据的 key 上,并直接输出
        public static class Reduce extends Reducer<Text, NullWritable, Text, NullWritable>{
        //实现 reduce 函数
        public void reduce(Text key, Iterable<NullWritable> values, Context context)
throws IOException, InterruptedException{
            context.write(key,NullWritable.get());
        }
        }
        public static void main(String[] args) throws IOException, ClassNotFoundException,
InterruptedException{
        Configuration conf = new Configuration();
        FindHDFSText find = new FindHDFSText();
        conf.set("dfs.client.use.datanode.hostname", "true");
        System.out.println("start");
        @SuppressWarnings("deprecation")
        Job job = new Job(conf,"filter");
        job.setJarByClass(Filter.class);
        //设置 Map、Combine 和 Reduce 处理类
        job.setMapperClass(Map.class);
        job.setReducerClass(Reduce.class);
        job.setOutputKeyClass(Text.class);
        job.setOutputValueClass(NullWritable.class);
        job.setInputFormatClass(TextInputFormat.class);
        job.setOutputFormatClass(TextOutputFormat.class);
        Path in = new Path("hdfs:// 192.168.1.100:9000/mymapreduce2/input/user");
        Path out = new Path("hdfs:// 192.168.1.100:9000/data/output");
        Path path = new Path("hdfs:// 192.168.31.70:9000/mymapreduce2/output");
        FileSystem fileSystem = path.getFileSystem(conf);// 根据 path 找到这个文件
        if (fileSystem.exists(path)) {
          fileSystem.delete(path, true);//true 表示就算 output 有内容,也会一并删除
        }
        FileInputFormat.addInputPath(job,in);
        FileOutputFormat.setOutputPath(job,out);

        System.exit(job.waitForCompletion(true)? 0:1);
        }
    }
```

在 Filter 类文件中,右击“Run As→Run on Hadoop”选项,将 MapReduce 任务提交到 Hadoop 中。

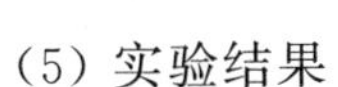

(5) 实验结果

在 HDFS 中/mymapreduce2/output 查看实验结果。

```
[hadoop@slave hadoop]$ hadoop dfs -ls /mymapreduce2/output
-rw-r--r-- 1 hadoop supergroup 0 2019-08-06 17:00 / mymapreduce2/output/_SUCCESS
-rw-r--r-- 1 hadoop supergroup 60 2019-08-06 17: 00 /mymapreduce2/output/part-r-00000
[hadoop@slave hadoop]$ hadoop dfs -cat /mymapreduce2/output/part-r-00000
100371
101258
101450
101487
101569
102051
102310
102312
102317
102319
103045
103054
103056
103084
103090
103091
103093
103179
103180
103182
103216
110068
110073
110565
```

3. 实验 3:二次排序

(1) 实验目的

基于 MapReduce 思想,编写二次排序程序。

(2) 实验要求

① 理解 MapReduce 编程思想。

② 编写 MapReduce 版二次排序程序。

③ 二次排序就是首先按照第一字段排序,然后对第一字段相同的行按照第二字段排序,注意不能破坏第一次排序的结果。

④ 知悉执行过程。

(3) 实验原理

首先我们需要知悉 MapReduce 的处理流程，其中，MapReduce 默认以 key 排序[11]。我们可以通过自定义 key，如数据格式<(key, value), value>，只需在分区的时候把 key 相同的值划分到同一个 Reduce 中，在 Reduce 中，只需要对组合 key 排序和分组即可实现。可以把实现过程分为以下步骤。

① 自定义组合键(key)。由 key 和要排序的 value 组成，并实现 WritableComparable 接口和 compareTo()方法的比较策略。

② 自定义分区函数类。将相同 key 分配到同一个 Reduce 中，要继承 Partitioner，重写 getPartition 函数。自定义分区函数类 FirstPartitioner，这是 key 的第一次比较，完成对所有 key 的排序。在 job 中使用 setPartitionerClass()方法设置 Partitioner job. setPartitionerClass (FirstPartitioner. Class)。

③ 自定义分组比较器。要继承 WritableComparator，必须有一个构造函数，并且重载以下方法：public int compare(WritableComparable w1, WritableComparable w2)。

④ 自定义排序比较器。这是 key 的第二次比较，对所有的 key 进行排序，即同时完成 IntPair 中的 first 和 second 排序。该类是一个比较器，可以通过两种方式实现，一是在 Job 中通过 setGroupingComparatorClass()方法来设置分组类，二是实现接口 RawComparator。在 Job 中，两种方式都可以通过 setSortComparatorClass()方法来设置 key 的比较类。继承 WritableComparator，必须有一个构造函数，并且重载以下方法：public int compare (WritableComparable w1, WritableComparable w2)。

(4) 实验步骤

① 开启 Hadoop。

```
[hadoop@master ~]$ cd /modules/hadoop/
[hadoop@master hadoop]$ sbin/start-all.sh
```

② 在 HDFS 上新建文件目录。

```
[hadoop@slave hadoop]$ hdfs dfs -mkdir -p /mymapreduce3/input
[hadoop@slave hadoop]$ hdfs dfs -mkdir -p /mymapreduce3/output
```

③ 上传数据文件到 HDFS。在 Linux 本地/data 目录下创建 data. txt 文件，并导入 HDFS 的/mymapreduce3/input 目录中。

```
[hadoop@slave hadoop]$ bin/hadoop dfs -put /data/data.txt /mymapreduce3/input
```

④ 用 Java 编写程序。Java 代码主要分为 4 个部分：自定义组合键(key)，自定义分区函数类，实现 Map，实现 Reduce。

自定义组合键(key)的代码：

```
public static class IntPair implements WritableComparable<IntPair>
{
    int first;          //第一个成员变量
    int second;         //第二个成员变量

    public void set(int left, int right)
```

```
{
    first = left;
    second = right;
}
public int getFirst()
{
    return first;
}
public int getSecond()
{
    return second;
}
@Override
//反序列化,从流中的二进制转换成 IntPair
public void readFields(DataInput in) throws IOException
{
    first = in.readInt();
    second = in.readInt();
}
@Override
//序列化,将 IntPair 转化成使用流传送的二进制
public void write(DataOutput out) throws IOException
{
    out.writeInt(first);
    out.writeInt(second);
}
@Override
//key 的比较
public int compareTo(IntPair o)
{
    if (first != o.first)
    {
        return first < o.first ? 1 : -1;
    }
    else if (second != o.second)
    {
        return second < o.second ? -1 : 1;
    }
    else
    {
        return 0;
```

```
        }
    }
    @Override
    public int hashCode()
    {
        return first * 157 + second;
    }
    @Override
    public boolean equals(Object right)
    {
        if (right == null)
            return false;
        if (this == right)
            return true;
        if (right instanceof IntPair)
        {
            IntPair r = (IntPair) right;
            return r.first == first && r.second == second;
        }
        else
        {
            return false;
        }
    }
}
```

自定义中需要实现接口 WritableComparable,因为是可序列的并且可比较的,并重载方法。该类中包含以下几种方法:反序列化、序列化、key 的比较。

分区函数类代码:

```
public static class FirstPartitioner extends Partitioner<IntPair, IntWritable>
    {
        @Override
        public int getPartition(IntPair key, IntWritable value,int numPartitions)
        {
            return Math.abs(key.getFirst() % numPartitions);
        }
    }
```

对 key 进行分区，是为实现第一次排序。

分组函数类代码：

```
public static class GroupingComparator extends WritableComparator
    {
        protected GroupingComparator()
        {
            super(IntPair.class, true);
        }
        @Override
        public int compare(WritableComparable w1, WritableComparable w2)
        {
            IntPair ip1 = (IntPair) w1;
            IntPair ip2 = (IntPair) w2;
            int l = ip1.getFirst();
            int r = ip2.getFirst();
            return l == r ? 0 : (l < r ? -1 : 1);
        }
    }
```

分组函数类直接比较复合键的第一个值即可，这是一个比较器，需要继承 WritableComparator。

Map 代码：

```
public static class Map extends Mapper<LongWritable, Text, IntPair, IntWritable> {

    private final IntPair key = new IntPair();
    private final IntWritable value = new IntWritable();

    @Override
    public void map(LongWritable inKey, Text inValue, Context context) throws
IOException, InterruptedException {
    StringTokenizer itr = new StringTokenizer(inValue.toString());
    int left = 0;
    int right = 0;
    if (itr.hasMoreTokens()) {
        left = Integer.parseInt(itr.nextToken());
        if (itr.hasMoreTokens()) {
            right = Integer.parseInt(itr.nextToken());
        }
```

```
            key.set(left, right);
            value.set(right);
            context.write(key, value);
        }
    }
}
```

Reduce 代码：

```
public static class Reduce extends Reducer<IntPair, IntWritable, Text, IntWritable>{
        private final Text left = new Text();
        private static final Text SEPARATOR = new Text("------------------------------------");

        public void reduce(IntPair key, Iterable<IntWritable> values,Context
context) throws IOException, InterruptedException {
            context.write(SEPARATOR, null);
            left.set(Integer.toString(key.getFirst()));
            System.out.println(left);
            for (IntWritable val : values)
            {
            context.write(left, val);
            //System.out.println(val);
            }
        }
}
```

完整代码：

```
package mapreduce;
import java.io.DataInput;
import java.io.DataOutput;
import java.io.IOException;
import java.util.StringTokenizer;
import org.apache.hadoop.conf.Configuration;
import org.apache.hadoop.fs.Path;
import org.apache.hadoop.io.IntWritable;
import org.apache.hadoop.io.LongWritable;
import org.apache.hadoop.io.Text;
import org.apache.hadoop.io.WritableComparable;
import org.apache.hadoop.io.WritableComparator;
import org.apache.hadoop.mapreduce.Job;
```

```
import org.apache.hadoop.mapreduce.Mapper;
import org.apache.hadoop.mapreduce.Partitioner;
import org.apache.hadoop.mapreduce.Reducer;
import org.apache.hadoop.mapreduce.lib.input.FileInputFormat;
import org.apache.hadoop.mapreduce.lib.input.TextInputFormat;
import org.apache.hadoop.mapreduce.lib.output.FileOutputFormat;
import org.apache.hadoop.mapreduce.lib.output.TextOutputFormat;
public class MySecondarySort
{

    public static class IntPair implements WritableComparable<IntPair>
    {
    int first;
    int second;

    public void set(int left, int right)
    {
        first = left;
        second = right;
    }
    public int getFirst()
    {
        return first;
    }
    public int getSecond()
    {
        return second;
    }
    @Override
    public void readFields(DataInput in) throws IOException
    {
        first = in.readInt();
        second = in.readInt();
    }
    @Override
    public void write(DataOutput out) throws IOException
    {
        out.writeInt(first);
        out.writeInt(second);
    }
```

```
@Override
public int compareTo(IntPair o)
{
    if (first != o.first)
    {
        return first < o.first ? 1 : -1;
    }
    else if (second != o.second)
    {
        return second < o.second ? -1 : 1;
    }
    else
    {
        return 0;
    }
}
}

public static class FirstPartitioner extends Partitioner<IntPair, IntWritable> {
@Override
public int getPartition(IntPair key, IntWritable value,int numPartitions)
{
    return Math.abs(key.getFirst() % numPartitions);
}
}
//如果不添加这个类,默认第一列和第二列都是升序排序的
//这个类的作用就是使第一列升序排序,第二列降序排序
public static class GroupingComparator extends WritableComparator
{ //无参构造器必须加上,否则报错
    protected GroupingComparator()
{
    super(IntPair.class, true);
}
@Override
public int compare(WritableComparable w1, WritableComparable w2)
```

```
{
    IntPair ip1 = (IntPair) w1;
    IntPair ip2 = (IntPair) w2;
    int l = ip1.getFirst();
    int r = ip2.getFirst();
    return l == r ? 0 : (l < r ? -1 : 1);
}
}

public static class Map extends Mapper<LongWritable, Text, IntPair, IntWritable> {

private final IntPair key = new IntPair();
private final IntWritable value = new IntWritable();

@Override
public void map(LongWritable inKey, Text inValue, Context context) throws
IOException, InterruptedException {
    StringTokenizer itr = new StringTokenizer(inValue.toString());
    int left = 0;
    int right = 0;
    if (itr.hasMoreTokens()) {
        left = Integer.parseInt(itr.nextToken());
        if (itr.hasMoreTokens()) {
            right = Integer.parseInt(itr.nextToken());
        }
        key.set(left, right);
        value.set(right);
        context.write(key, value);
    }
}
}
public static class Reduce extends Reducer<IntPair, IntWritable, Text, IntWritable>
{
private final Text left = new Text();
private static final Text SEPARATOR = new Text("------------------------------------");

public void reduce(IntPair key, Iterable<IntWritable> values, Context
context) throws IOException, InterruptedException
```

```
        {
            context.write(SEPARATOR, null);
            left.set(Integer.toString(key.getFirst()));
            System.out.println(left);
            for (IntWritable val : values)
            {
                context.write(left, val);
                //System.out.println(val);
            }
        }
        }
        public static void main(String[] args) throws IOException,
InterruptedException, ClassNotFoundException
        {

            Configuration conf = new Configuration();
            Job job = new Job(conf, "secondarysort");
            job.setJarByClass(SecondarySort.class);
            job.setMapperClass(Map.class);
            job.setReducerClass(Reduce.class);
            job.setPartitionerClass(FirstPartitioner.class);
            job.setGroupingComparatorClass(GroupingComparator.class);
            job.setMapOutputKeyClass(IntPair.class);
            job.setMapOutputValueClass(IntWritable.class);
            job.setOutputKeyClass(Text.class);
            job.setOutputValueClass(IntWritable.class);
            job.setInputFormatClass(TextInputFormat.class);
            job.setOutputFormatClass(TextOutputFormat.class);
            String[] otherArgs = new String[2];
            Args[0] = "hdfs://192.168.1.100:9000/mymapreduce3/input/data.txt";
            Args[1] = "hdfs://192.168.1.100:9000/mymapreduce3/output";
            FileInputFormat.setInputPaths(job, new Path(Args[0]));
            FileOutputFormat.setOutputPath(job, new Path(Args[1]));
            System.exit(job.waitForCompletion(true) ? 0 : 1);
        }
    }
```

(5) 实验结果

① 输入数据/data/data.txt。

```
3    119
1    23
4    6
1    22
4    1
1    24
4    2
3    10
```

② 通过以下命令对 HDFS 里的文件/mymapreduce2/output/part-r-00000 查看实验结果。

```
[hadoop@slave hadoop]$ hadoop dfs -cat /mymapreduce2/output/part-r-00000
1    22
1    23
1    24
----------------------------
4    1
4    2
4    6
----------------------------
3    10
3    119
```

4. 实验4:Map 端 Join

(1) 实验目的

MapReduce 提供了表连接操作,其中包括 Map 端 Join、Reduce 端 Join 还有单表 Join 三种方式,本实验讨论的是 Map 端 Join 操作程序。

(2) 实验要求

① 理解 MapReduce 编程思想。

② 编写 MapReduce 版 Map 端 Join 操作程序。

③ 熟悉 Map 端 Join 的解决方案。

(3) 实验原理

① Map 端的使用场景:一张表数据十分小、一张表数据很大。

a. 主表:order.txt。从表:thing.txt。

b. 主表正常进行 map 处理,从表以缓存文件的形式进行读入,并加入 job 中,主表数据在 Map 中处理之后,缺少从表数据的完整性,就需要对存放了主表数据的完整对象,设置缺失的从表数据。

c. 根据 Mapper 类中 setup 方法只执行一次的特性,对缓存文件进行加载,将解析后的储存了商品信息的对象以<商品 ID,商品对象>形式加入 Map 中,方便我们使用。这样在 Map

中，我们通过 map. get(key)就能取到当前这个订单所对应的商品，进行赋值和写入即实现。

② Map 端 Join 的执行流程。

a. 首先将小表文件放到该作业的 DistributedCache 中，然后从 DistributeCache 中取出该小表进行 Join 连接的<key ,value>键值对，将其解释分割放到内存中。

b. 重写自定义 XxMapper 类，并实现 Mapper 类的 setup()方法，因为这个方法是先于 map 方法执行的，将较小表先读入一个 HashMap 中。

c. 重写 map 函数，一行行读入大表的数据，与 HashMap 中的内容一一进行比较，若 key 相同，则对数据进行格式化处理，然后直接输出。

d. map 函数输出的<key，value>键值对首先经过一个 Shuffle 把 key 值相同的所有 value 放到一个迭代器中形成 values，然后将<key，values>键值对传递给 reduce 函数，reduce 函数输入的 key 直接复制给输出的 key，输入的 values 通过增强版 for 循环遍历逐一输出，循环的次数决定了<key，value>输出的次数。

（4）实验步骤

① 开启 Hadoop。

```
[hadoop@master ~]$ cd /modules/hadoop/
[hadoop@master hadoop]$ sbin/start-all.sh
```

② 在 HDFS 上新建文件目录。

```
[hadoop@slave hadoop]$ hdfs dfs -mkdir -p /mymapreduce4/input
[hadoop@slave hadoop]$ hdfs dfs -mkdir -p /mymapreduce4/output
```

③ 准备实验数据：order. txt 和 thing. txt。

order. txt：

```
1101  1   2018-12-12 14:50:21
1102  2   2018-12-15 12:38:21
1103  1   2018-10-15 09:47:07
1104  3   2018-11-15 18:58:41
```

thing. txt：

```
1  huawei
2  xiaomi
3  oneplus
```

现在要通过写 Map 端 Join 程序得到 order. txt 表中的第二个字段和 thing. txt 表中第一个字段相同的数据的 Join 结果：

```
1    1101    2018-12-12 14:50:21    huawei
1    1103    2018-10-15 09:47:07    huawei
2    1102    2018-12-15 12:38:21    xiaomi
3    1103    2018-11-15 18:58:41    oneplus
```

④ 上传数据文件到 HDFS。在 Linux 本地/data 目录下创建 order. txt 和 thing. txt 文件，并导入 HDFS 的/mymapreduce4/input 目录中。

```
[hadoop@slave hadoop]$ bin/hadoop dfs -put /data/order.txt /mymapreduce4/input
[hadoop@slave hadoop]$ bin/hadoop dfs -put /data/thing.txt /mymapreduce4/input
```

⑤ 编写程序代码。MapReduce 的 Java 代码分为两个部分:实现 Mapper,实现 Reduce。

Mapper 代码:

```
public static class Mapper extends Mapper<Object, Text, Text, Text>{
    private Map<String, String> dict = new HashMap<>();

    @Override
    protected void setup(Context context) throws IOException,InterruptedException {
    String fileName = context.getLocalCacheFiles()[0].getName();
    System.out.println(fileName);
    BufferedReader reader = new BufferedReader(new FileReader(fileName));
    String codeandname = null;
    while (null != ( codeandname = reader.readLine() ) ) {
        String str[] = codeandname.split("\t");
        dict.put(str[1], str[0] + "\t" + str[2]);
    }
    reader.close();
    }
    @Override
    protected void map(Object key, Text value, Context context) throws IOException,
InterruptedException {
            String[] kv = value.toString().split("\t");
            if (dict.containsKey(kv[0])) {
                context.write(new Text(kv[0]), new Text(dict.get(kv[0]) + "\t" +
                                                        kv[1]));
            }
    }
}
```

Reduce 代码:

```
public static class Reducer extends Reducer<Text, Text, Text, Text>{
    @Override
    protected void reduce(Text key, Iterable<Text> values, Context context)
throws IOException, InterruptedException {
```

```
        for (Text text : values) {
            context.write(key, text);
        }
    }
}
```

完整代码：

```
package mapreduce;
import java.io.BufferedReader;
import java.io.FileReader;
import java.io.IOException;
import java.net.URI;
import java.net.URISyntaxException;
import java.util.HashMap;
import java.util.Map;
import org.apache.hadoop.fs.Path;
import org.apache.hadoop.io.Text;
import org.apache.hadoop.mapreduce.Job;
import org.apache.hadoop.mapreduce.Mapper;
import org.apache.hadoop.mapreduce.Reducer;
import org.apache.hadoop.mapreduce.lib.input.FileInputFormat;
import org.apache.hadoop.mapreduce.lib.output.FileOutputFormat;
public class MapJoin {

    public static class Mapper extends Mapper<Object, Text, Text, Text>{
        private Map<String, String> dict = new HashMap<>();

        @Override
        protected void setup(Context context) throws IOException, InterruptedException {
            String fileName = context.getLocalCacheFiles()[0].getName();
            //System.out.println(fileName);
            BufferedReader reader = new BufferedReader(new FileReader(fileName));
            String codeandname = null;
            while (null != ( codeandname = reader.readLine() ) ) {
                String str[] = codeandname.split("\t");
                dict.put(str[1], str[0] + "\t" + str[2]);
            }
```

```
            reader.close();
        }
        @Override
        protected void map(Object key, Text value, Context context)throws IOException,
InterruptedException {
            String[] kv = value.toString().split("\t");
            if (dict.containsKey(kv[0])) {
                context.write(new Text(kv[0]), new Text(dict.get(kv[0]) + "\t" +
                                                    kv[1]));
            }
        }
    }
    public static class Reducer extends Reducer<Text, Text, Text, Text>{
        @Override
        protected void reduce(Text key, Iterable<Text> values, Context context) throws
IOException, InterruptedException {
            for (Text text : values) {
            context.write(key, text);
            }
        }
    }

    public static void main(String[] args) throws ClassNotFoundException,
IOException, InterruptedException, URISyntaxException {
    Job job = Job.getInstance();
    job.setJobName("mapjoin");
    job.setJarByClass(MapJoin.class);

    job.setMapperClass(MyMapper.class);
    job.setReducerClass(MyReducer.class);

    job.setOutputKeyClass(Text.class);
    job.setOutputValueClass(Text.class);

    Path in = new Path("hdfs://192.168.1.100:9000/mymapreduce4/input/thing.
                txt");
    Path out = new Path("hdfs:// 192.168.1.100:9000/mymapreduce4/output");
    FileInputFormat.addInputPath(job, in);
```

```
        FileOutputFormat.setOutputPath(job, out);

        URI uri = new URI("hdfs:// 192.168.1.100:9000/mymapreduce4/input/order.
                txt");
        job.addCacheFile(uri);

        System.exit(job.waitForCompletion(true) ? 0 : 1);
    }
}
```

(5) 实验结果

在 HDFS 上查看结果。

```
[hadoop@slave hadoop]$ hadoop dfs -ls /mymapreduce4/output
-rw-r--r-- 3 hadoop supergroup 0 2019-08-08 19:20 / mymapreduce4/output/_SUCCESS
-rw-r--r-- 3 hadoop supergroup 182 2019-08-08 19: 20 /mymapreduce4/output/part-
r-00000
[hadoop@slave hadoop]$ hadoop dfs -cat /mymapreduce4/output/part-r-00000
1     1101     2018-12-12 14:50:21     huawei
1     1103     2018-10-15 09:47:07     huawei
2     1102     2018-12-15 12:38:21     xiaomi
3     1103     2018-11-15 18:58:41     oneplus
```

5. 实验 5:MapReduce 自定义格式输出

(1) 实验目的

基于 MapReduce 思想,编写 MapReduce 自定义格式输出程序。

(2) 实验要求

理解 MapReduce 编程思想,编写 MapReduce 版 MapReduce 自定义输出格式程序,有助于理解和使用 MapReduce 输出格式类。

(3) 实验原理

① 输出格式。OutputFormat 负责把 Reducer 处理完成的 key-value 写到本地磁盘或 HDFS 上,默认计算结果会以 part-000 命名输出成多个文件,并且输出的文件数量与 Reduce 数量一致[12]。提供给 OutputCollector 的键值对会被写到输出文件中,写入的方式由输出格式控制。MapReduce 提供多种输出格式,用户可以灵活设置输出的路径、文件名、输出格式等。可以通过具体 MapReduce 作业的 JobConf 对象的 setOutputFormat()方法来设置具体用到的输出格式:

TextOutputFormat:默认的输出格式,以"key \t value"的方式输出行。

SequenceFileOutputFormat:序列化文件输出,将 key 和 value 以 sequencefile 格式输出。

NullOutputFormat:忽略收到的数据,不输出任何数据。

SequenceFileAsBinaryOutputFormat:与 SequenceFileAsBinaryInputFormat 相对应,将键值对当作二进制数据写入一个顺序文件。

MultiOutputs：可以把输出数据输送到不同的目录。

Hadoop 提供了一些 OutputFormat 实例用于写入文件，基本的（默认的）实例是 TextOutputFormat，它会以一行一个键值对的方式把数据写入一个文本文件里。这样后面的 MapReduce 任务就可以通过 KeyValueInputFormat 类简单地重新读取所需的输入数据，而且也方便阅读。还有一个更适合在 MapReduce 作业间使用的中间格式，那就是 SequenceFileOutputFormat，它可以快速地序列化任意的数据类型到文件中，对应地，SequenceFileInputFormat 则会把文件反序列化为相同的类型并提交为下一个 Mapper 的输入数据，方式和前一个 Reducer 的生成方式一样。NullOutputFormat 不会生成输出文件并会丢弃任何通过 OutputCollector 传递给它的键值对，如果用户要在 reduce()方法中显式地写自己的输出文件并且不想 Hadoop 框架输出额外的空输出文件，那么这个类是很有用的。

② 自定义输出数据格式。支持多个输出，步骤如下所述。

a. 自定义的类要继承 OutputFormat 类，一般选择继承 FileOutputFormat 即可。

b. 实现 getRecordWriter 方法，返回一个 RecordWriter 类型。

c. 自定义类继承 RecordWriter，实现 write 方法。

（4）实验步骤

① 开启 Hadoop。

```
[hadoop@master ~]$ cd /modules/hadoop/
[hadoop@master hadoop]$ sbin/start-all.sh
```

② 在 HDFS 上新建文件目录。

```
[hadoop@slave hadoop]$ hdfs dfs -mkdir -p /mymapreduce5/input
[hadoop@slave hadoop]$ hdfs dfs -mkdir -p /mymapreduce5/output
```

③ 准备实验数据。

catetory.txt：

```
732   奢侈品     b   1
795   箱包       2   1
896   化妆品     3   1
897   家电       4   1
721   有机食品   3   0
722   蔬菜水果   3   0
723   肉禽蛋奶   4   0
724   深海水产   5   0
725   地方特产   6   0
726   进口食品   7   0
```

需要实现的是把 0 和 1 标记的数据分别放入两个文件中，并且以该字段来命名文件，结果输出 0.txt 和 1.txt 两个文件。

④ 上传数据文件到 HDFS。在 Linux 本地/data 目录下创建 category.txt 文件，并导入 HDFS 的/mymapreduce5/input 目录中。

```
[hadoop@slave hadoop]$ bin/hadoop dfs -put /data/category.txt /mymapreduce5/input
```

⑤ 编写程序代码。自定义的 FileRecordWriter 类命名为 DiyMultipleOutputFormat，该类继承了 FileRecordWriter 类，并且其中主要包含三部分：类中的 getRecordWriter、getTaskOutputPath、generateFileNameForKeyValue 方法和两个内部类 LineRecordWriter、MutiRecordWriter。

方法代码：

```
private MultiRecordWriter writer = null;
// getRecordWriter()方法判断该类实例是否存在，若不存在则创建一个实例
public RecordWriter < K, V > getRecordWriter ( TaskAttemptContext job) throws IOException{
        if(writer == null){
            writer = new MultiRecordWriter(job,getTaskOutputPath(job));
        }
        return writer;
}
//获取工作任务的输出路径
private Path getTaskOutputPath(TaskAttemptContext conf) throws IOException{
        Path workPath = null;
        OutputCommitter committer = super.getOutputCommitter(conf);
        if(committer instanceof FileOutputCommitter){
            workPath = ((FileOutputCommitter) committer).getWorkPath();
        }else{
            Path outputPath = super.getOutputPath(conf);
            if(outputPath == null){
                throw new IOException("Undefined job output-path");
            }
            workPath = outputPath;
        }
        return workPath;
}
//通过 key、value 和 conf 三个参数确定 key/value 输出的文件名
protected abstract String generateFileNameForKayValue ( K key, V value, Configuration conf);
```

LineRecordWriter 类主要是在 < key, value > 输出时定义它的输出格式。LineRecordWriter 类代码：

```
protected static class LineRecordWriter<K,V> extends RecordWriter<K, V> {
    private static final String utf8 = "UTF-8";
    private static final byte[] newline;
    private PrintWriter tt;
    static {
        try {
            newline = "\n".getBytes(utf8);
            } catch (UnsupportedEncodingException uee) {
                throw new IllegalArgumentException("can't find " + utf8 +
                                                    "encoding");
            }
    }

    protected DataOutputStream out;
    private final byte[] keyValueSeparator;

    public LineRecordWriter(DataOutputStream out, String keyValueSeparator) {
        this.out = out;
        try {
            this.keyValueSeparator = keyValueSeparator.getBytes(utf8);
            } catch (UnsupportedEncodingException uee) {
            throw new IllegalArgumentException("can't find " + utf8 + " encoding");
            }
    }

    public LineRecordWriter(DataOutputStream out) {
            this(out, ":");
    }
    private void writeObject(Object o) throws IOException {
        if (o instanceof Text) {
          Text to = (Text) o;
          out.write(to.getBytes(), 0, to.getLength());
        } else {
          out.write(o.toString().getBytes(utf8));
        }
    }

    public synchronized void write(K key, V value) throws IOException {
```

```
        boolean nullKey = key == null || key instanceof NullWritable;
        boolean nullValue = value == null || value instanceof NullWritable;
        if (nullKey && nullValue) {//
            return;
        }
        if (! nullKey) {
            writeObject(key);
        }
        if (! (nullKey || nullValue)) {
            out.write(keyValueSeparator);
        }
        if (! nullValue) {
            writeObject(value);
        }
        out.write(newline);

    }
    public synchronized void close(TaskAttemptContext context) throws IOException {
        out.close();
    }
}
```

MultiRecordWriter 类代码：

```
public class MultiRecordWriter extends RecordWriter<K,V>{
        private HashMap<String,RecordWriter<K,V> >recordWriters = null;
        private TaskAttemptContext job = null;
        private Path workPath = null;
        public MultiRecordWriter(TaskAttemptContext job,Path workPath){
            super();
            this.job = job;
            this.workPath = workPath;
            recordWriters = new HashMap<String,RecordWriter<K,V>>();

        }
        // close()方法关闭输出文件的数据流
        public void close(TaskAttemptContext context) throws IOException, Inter-
ruptedException{
```

```
            Iterator<RecordWriter<K,V>> values = this.recordWriters.values().
                                                            iterator();
            while(values.hasNext()){
                values.next().close(context);
            }
            this.recordWriters.clear();
        }
        // write()方法得到输出的文件 0.txt 和 1.txt,并将两文件写到 HDFS 上
        public void write(K key,V value) throws IOException, InterruptedException{
            String baseName = generateFileNameForKayValue(key ,value,job.
                                                            getConfiguration());
            RecordWriter<K,V> rw = this.recordWriters.get(baseName);
            if(rw == null){
                rw = getBaseRecordWriter(job,baseName);
                this.recordWriters.put(baseName,rw);
            }
            rw.write(key, value);
        }


         private RecordWriter<K,V> getBaseRecordWriter(TaskAttemptContext job,
String baseName)throws IOException,InterruptedException{
            Configuration conf = job.getConfiguration();
            boolean isCompressed = getCompressOutput(job);
            String keyValueSeparator = ":";
            RecordWriter<K,V> recordWriter = null;
            if(isCompressed){
                            Class <? extends CompressionCodec > codecClass =
getOutputCompressorClass(job,(Class<? extends CompressionCodec>) GzipCodec.
                        class);
                CompressionCodec codec = ReflectionUtils.newInstance(codecClass,conf);
                Path file = new Path(workPath,baseName + codec.getDefaultExtension());
                FSDataOutputStream fileOut = file.getFileSystem(conf).create
                                                (file,false);
                recordWriter = new LineRecordWriter<K,V>(new DataOutputStream(codec.
                            createOutputStream(fileOut)),keyValueSeparator);
```

```
            }else{
                Path file = new Path(workPath,baseName);
                FSDataOutputStream fileOut = file.getFileSystem(conf).create
                                        (file,false);
                recordWriter = new LineRecordWriter<K,V>(fileOut,keyValue-
                            Separator);
            }
            return recordWriter;
        }
    }
```

编写自定义输出格式类 DiyMultipleOutputFormat。DiyMultipleOutputFormat 完整代码：

```
package mapreduce;
import java.io.DataOutputStream;
import java.io.IOException;
import java.io.PrintWriter;
import java.io.UnsupportedEncodingException;
import java.util.HashMap;
import java.util.Iterator;
import org.apache.hadoop.conf.Configuration;
import org.apache.hadoop.fs.FSDataOutputStream;
import org.apache.hadoop.fs.Path;
import org.apache.hadoop.io.NullWritable;
import org.apache.hadoop.io.Text;
import org.apache.hadoop.io.Writable;
import org.apache.hadoop.io.WritableComparable;
import org.apache.hadoop.io.compress.CompressionCodec;
import org.apache.hadoop.io.compress.GzipCodec;
import org.apache.hadoop.mapreduce.OutputCommitter;
import org.apache.hadoop.mapreduce.RecordWriter;
import org.apache.hadoop.mapreduce.TaskAttemptContext;
import org.apache.hadoop.mapreduce.lib.output.FileOutputCommitter;
import org.apache.hadoop.mapreduce.lib.output.FileOutputFormat;
import org.apache.hadoop.util.ReflectionUtils;
public abstract class DiyMultipleOutputFormat <K extends WritableComparable<? >,
V extends Writable>extends FileOutputFormat<K,V>{
    private MultiRecordWriter writer = null;
```

```
    public RecordWriter<K,V> getRecordWriter(TaskAttemptContext job) throws
IOException{
        if(writer == null){
            writer = new MultiRecordWriter(job,getTaskOutputPath(job));
        }
        return writer;
    }
    private Path getTaskOutputPath(TaskAttemptContext conf) throws IOException{
        Path workPath = null;
        OutputCommitter committer = super.getOutputCommitter(conf);
        if(committer instanceof FileOutputCommitter){
            workPath = ((FileOutputCommitter) committer).getWorkPath();
        }else{
            Path outputPath = super.getOutputPath(conf);
            if(outputPath == null){
                throw new IOException("Undefined job output-path");
            }
            workPath = outputPath;
        }
        return workPath;
    }
    protected abstract String generateFileNameForKayValue(K key,V value,Configuration
conf);
    protected static class LineRecordWriter<K,V> extends RecordWriter<K, V> {
        private static final String utf8 = "UTF-8";
        private static final byte[] newline;
        private PrintWriter tt;
        static {
          try {
              newline = "\n".getBytes(utf8);
          } catch (UnsupportedEncodingException uee) {
              throw new IllegalArgumentException("can't find " + utf8 + " encoding");
          }
        }

        protected DataOutputStream out;
        private final byte[] keyValueSeparator;
```

```
    public LineRecordWriter(DataOutputStream out, String keyValueSeparator) {
        this.out = out;
        try {
            this.keyValueSeparator = keyValueSeparator.getBytes(utf8);
        } catch (UnsupportedEncodingException uee) {
            throw new IllegalArgumentException("can't find " + utf8 + "encoding");
        }
    }

    public LineRecordWriter(DataOutputStream out) {
        this(out, ":");
    }
    private void writeObject(Object o) throws IOException {
        if (o instanceof Text) {
            Text to = (Text) o;
            out.write(to.getBytes(), 0, to.getLength());
        } else {
            out.write(o.toString().getBytes(utf8));
        }
    }

    public synchronized void write(K key, V value) throws IOException {
        boolean nullKey = key == null || key instanceof NullWritable;
        boolean nullValue = value == null || value instanceof NullWritable;
        if (nullKey && nullValue) {
            return;
        }
        if (! nullKey) {
            writeObject(key);
        }
        if (! (nullKey || nullValue)) {
            out.write(keyValueSeparator);
        }
        if (! nullValue) {
```

```
            writeObject(value);
        }
        out.write(newline);

    }

    public synchronized
    void close(TaskAttemptContext context) throws IOException {
        out.close();
    }
}
public class MultiRecordWriter extends RecordWriter<K,V>{
    private HashMap<String,RecordWriter<K,V> >recordWriters = null;
    private TaskAttemptContext job = null;
    private Path workPath = null;
    public MultiRecordWriter(TaskAttemptContext job,Path workPath){
        super();
        this.job = job;
        this.workPath = workPath;
        recordWriters = new HashMap<String,RecordWriter<K,V>>();

    }
    public void close(TaskAttemptContext context) throws IOException,
InterruptedException{
        Iterator<RecordWriter<K,V>> values = this.recordWriters.values().
                                                    iterator();
        while(values.hasNext()){
            values.next().close(context);
        }
        this.recordWriters.clear();
    }
    public void write(K key,V value) throws IOException, InterruptedException{
        String baseName = generateFileNameForKayValue(key ,value,job.
                        getConfiguration());
        RecordWriter<K,V> rw = this.recordWriters.get(baseName);
        if(rw == null){
            rw = getBaseRecordWriter(job,baseName);
            this.recordWriters.put(baseName,rw);
        }
```

```
            rw.write(key, value);
        }


        private RecordWriter<K,V> getBaseRecordWriter(TaskAttemptContext job,
String baseName)throws IOException,InterruptedException{
            Configuration conf = job.getConfiguration();
            boolean isCompressed = getCompressOutput(job);
            String keyValueSeparator = ":";
            RecordWriter<K,V> recordWriter = null;
            if(isCompressed){
            Class<? extends CompressionCodec> codecClass = getOutputCompress-
orClass(job,(Class<? extends CompressionCodec>) GzipCodec.class);
            CompressionCodec codec = ReflectionUtils.newInstance(codecClass,conf);
            Path file = new Path(workPath,baseName + codec.getDefaultExtension());
            FSDataOutputStream fileOut = file.getFileSystem(conf).create(file,false);
            recordWriter = new LineRecordWriter<K,V>(new DataOutputStream
                    (codec.createOutputStream(fileOut)),keyValueSeparator);
            }else{
                Path file = new Path(workPath,baseName);
                FSDataOutputStream fileOut = file.getFileSystem(conf).create
                                    (file,false);
                recordWriter  = new LineRecordWriter<K,V>(fileOut,keyValue-
                        Separator);
            }
            return recordWriter;
        }
    }
}
```

编写一个测试类进行测试，测试类的完整代码：

```
package mapreduce;
import java.io.IOException;
import org.apache.hadoop.conf.Configuration;
import org.apache.hadoop.fs.Path;
import org.apache.hadoop.io.Text;
import org.apache.hadoop.mapreduce.Job;
```

```
    import org.apache.hadoop.mapreduce.Mapper;
    import org.apache.hadoop.mapreduce.Reducer;
    import org.apache.hadoop.mapreduce.lib.input.FileInputFormat;
    import org.apache.hadoop.mapreduce.lib.output.FileOutputFormat;
    public class FileOutputMRTest {
        public static class TokenizerMapper extends Mapper<Object,Text,Text,Text>{
            private Text val = new Text();
            public void map(Object key,Text value,Context context)throws IOException,
InterruptedException{
                String str[] = value.toString().split("\t");
                val.set(str[0] + " " + str[1] + " " + str[2]);
                context.write(new Text(str[3]), val);
            }
        }
        public static class IntSumReducer extends Reducer<Text,Text,Text,Text>{
            public void reduce(Text key,Iterable<Text> values,Context context)
throws IOException,InterruptedException{
                for(Text val:values){
                    context.write(key,val);
                }
            }
        }
        public static class MyMultipleOutputFormat extends DiyMultipleOutputFormat
<Text,Text>{
        protected String generateFileNameForKayValue(Text key,Text value,Configura-
tion conf){
            return key + ".txt";
          }
        }
        public static void main(String[] args) throws IOException, InterruptedException,
ClassNotFoundException{
        Configuration conf = new Configuration();
        Job job = new Job(conf,"FileOutputMR");
        job.setJarByClass(FileOutputMR.class);
        job.setMapperClass(TokenizerMapper.class);
        job.setCombinerClass(IntSumReducer.class);
        job.setReducerClass(IntSumReducer.class);
        job.setOutputKeyClass(Text.class);
```

```
        job.setOutputValueClass(Text.class);
        job.setOutputFormatClass(MyMultipleOutputFormat.class);
        FileInputFormat.addInputPath(job,new Path("hdfs://192.168.1.100:9000/
                                          mymapreduce5/input/category.
                                          txt"));
        FileOutputFormat.setOutputPath(job,new Path("hdfs:// 192.168.1.100:9000/
                                          mymapreduce5/output"));
        System.exit(job.waitForCompletion(true)? 0:1);
    }
}
```

(5) 实验结果

待执行完毕后，进入命令模式，在 HDFS 上的 mymapreduce5/output 中查看实验结果。

```
[hadoop@slave hadoop]$ hadoop dfs -ls /mymapreduce5/output
-rw-r--r-- 3 hadoop supergroup 0 2019-08-10 14:12 / mymapreduce5/output/_SUCCESS
-rw-r--r-- 3 hadoop supergroup 126 2019-08-10 14:12 /mymapreduce5/output/0.txt
-rw-r--r-- 3 hadoop supergroup 66 2019-08-10 14:12 /mymapreduce5/output/1.txt
[hadoop@slave hadoop]$ hadoop dfs -cat /mymapreduce5/output/0.txt
0:726   进口食品   7
0:725   地方特产   6
0:724   深海水产   5
0:723   肉禽蛋奶   4
0:722   蔬菜水果   3
0:721   有机食品   3
[hadoop@slave hadoop]$ hadoop dfs -cat /mymapreduce5/output/1.txt
1:897   家电       4
1:896   化妆品     3
1:895   箱包       2
1:732   奢侈品     b
```

2.5 实验作业

① 下载 Hadoop 安装包并在虚拟机上安装。

② 模拟实验内容部分。

2.6 扩展资料

在伪分布和完全分布两种模式下运行 MapReduce 示例 WordCount，便于为开展 MapReduce 的实验做好准备。

1. 实验1：伪分布模式

① 创建一个普通用户：hadoop。

```
[root@bigdata001 ~]$ user hadoop
[root@bigdata001 ~]$ passwd hadoop
```

② 向 hadoop 用户提供 sudo 权限。

```
[root@bigdata001 ~]$ vi /etc/sudoers
```

设置权限时，学习环境可以将 hadoop 用户的权限设置得大一些，但是生产环境一定要注意普通用户的权限限制。

```
root     ALL=(ALL)         ALL
hadoop ALL=(root) NOPASSWD:ALL
```

如果 root 用户无权修改 sudoers 文件，则首先手动为 root 用户添加写权限。

```
[root@bigdata001 ~]$ chmod u+w /etc/sudoers
```

③ 创建存放 Hadoop 文件的目录。

```
[hadoop@bigdata001 ~]$ sudo mkdir /apps/modules
```

指定 Hadoop 文件夹的所有者为 hadoop 用户，若存放 Hadoop 的目录的所有者不是 hadoop，之后 Hadoop 运行中可能会出现权限问题，因此将所有者改为 hadoop。

```
[hadoop@bigdata001 ~]$ sudo chown -R hadoop:hadoop /apps/modules
```

④ 解压 Hadoop 目录文件。

```
[hadoop@bigdata001 ~]$ cd /apps/modules[hadoop@bigdata1 hadoop-3.1.2]$
[root@bigdata1modules]$ tar -zxvf hadoop-3.1.2.tar.gz
```

⑤ 配置 Hadoop 环境变量。

```
[hadoop@bigdata1 hadoop-3.1.2]$ vi /etc/profile
```

追加配置：

```
export HADOOP_HOME="/apps/modules/hadoop-3.1.2"
export PATH=$HADOOP_HOME/bin:$HADOOP_HOME/sbin:$PATH
```

执行“source /etc/profile”使其配置生效。

⑥ 配置 hadoop-env.sh、mapred-env.sh、yarn-env.sh 文件的 JAVA_HOME 参数，修改 JAVA_HOME 路径并与/etc/profile 文件的 JAVA_HOME 路径保持一致。

```
[hadoop@bigdata001 hadoop-3.1.2]$ sudo vi etc/hadoop/hadoop-env.sh
```

Hadoop 还需要配置的文件有 core-site.xml、hdfs-site.xml、yarn-site.xml、mapred-site.xml。

⑦ 配置 core-site. xml。

```
<configuration>
 <property>
      <name>fs.defaultFS</name>
      <value>hdfs://bigdata1:9000</value>
  </property>
  <property>
       <name>hadoop.tmp.dir</name>
       <value>/apps/modules/tmp</value>
  </property>
</configuration>
```

hadoop. tmp. dir 配置的是 Hadoop 临时目录，HDFS 的 NameNode 数据默认存放在这个目录下，查看“＊-default. xml”等默认配置文件就可以看到很多依赖“＄{hadoop. tmp. dir}”的配置。

默认的 hadoop. tmp. dir 是/tmp/hadoop-＄{user. name}，此时 NameNode 会将 HDFS 的元数据存储在/tmp 目录下，如果操作系统重启，则系统会清空/tmp 目录下的内容，导致 NameNode 元数据丢失，这是个非常严重的问题，因此应该修改这个路径。

创建临时目录：

```
[hadoop@bigdata001 hadoop-3.1.2]$ sudo mkdir -p /apps/data/tmp
```

将临时目录的所有者修改为 hadoop：

```
[hadoop@bigdata001 hadoop-3.1.2]$ sudo chown -R hadoop:hadoop /apps/data/tmp
```

⑧ 配置 hdfs-site. xml。

```
<configuration>
<property>
     <name>dfs.replication</name>
     <value>1</value>
</property>
</configuration>
```

⑨ 配置 mapred-site. xml。

```
<configuration>
    <property>
           <name>mapreduce.framework.name</name>
           <value>yarn</value>
    </property>
    <property>
           <name>mapreduce.application.classpath</name>
           <value>
```

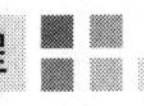

```
        /apps/modules/hadoop-3.1.2/etc/hadoop:/apps/modules/hadoop-3.1.2/share/hadoop/
common/lib/*:/apps/modules/hadoop-3.1.2/share/hadoop/common/*:/apps/modules/
hadoop-3.1.2/share/hadoop/hdfs:/apps/modules/hadoop-3.1.2/share/hadoop/hdfs/lib/
*:/apps/modules/hadoop-3.1.2/share/hadoop/hdfs/*:/apps/modules/hadoop-3.1.2/
share/hadoop/mapreduce/lib/*:/apps/modules/hadoop-3.1.2/share/hadoop/mapreduce/
*:/apps/modules/hadoop-3.1.2/share/hadoop/yarn:/apps/modules/hadoop-3.1.2/share/
hadoop/yarn/lib/*:/apps/modules/hadoop-3.1.2/share/hadoop/yarn/*
            </value>
        </property>
    </configuration>
```

上述 mapreduce.application.classpath 一开始没有配置，会导致使用 mapreduce 时报错：

```
Error: Could not find or load main class
org.apache.hadoop.mapreduce.v2.app.MRAppMaster
```

⑩ 配置 yarn-site.xml。

```
<configuration>
<property>
    <name>yarn.resourcemanager.hostname</name>
    <value>bigdata001</value>
</property>
<property>
    <name>yarn.nodemanager.aux-services</name>
    <value>mapreduce_shuffle</value>
</property>
<property>
    <name>yarn.log-aggregation-enable</name>
    <value>true</value>
</property>
<property>
<name>yarn.log-aggregation.retain-seconds</name>
    <value>106800</value>
</property>
</configuration>
```

yarn.nodemanager.aux-services 配置了 YARN 的默认混洗方式，选择为 MapReduce 的默认混洗算法。yarn.resourcemanager.hostname 指定了 resourcemanager 运行的节点。

⑪ 格式化 NameNode。

```
[hadoop@bigdata001 hadoop-3.1.2]$ hdfs namenode -format
```

如果出现图 2-2 所示的信息则说明格式化成功。

```
2017-12-26 23:19:37,515 INFO util.GSet: 0.25% max memory 1.7 GB = 4.4 MB
2017-12-26 23:19:37,515 INFO util.GSet: capacity      = 2^19 = 524288 entries
2017-12-26 23:19:37,525 INFO metrics.TopMetrics: NNTop conf: dfs.namenode.top.window.num.buckets = 10
2017-12-26 23:19:37,525 INFO metrics.TopMetrics: NNTop conf: dfs.namenode.top.num.users = 10
2017-12-26 23:19:37,525 INFO metrics.TopMetrics: NNTop conf: dfs.namenode.top.windows.minutes = 1,5,25
2017-12-26 23:19:37,530 INFO namenode.FSNamesystem: Retry cache on namenode is enabled
2017-12-26 23:19:37,530 INFO namenode.FSNamesystem: Retry cache will use 0.03 of total heap and retry cache entry expiry time is 600000 mi
2017-12-26 23:19:37,532 INFO util.GSet: Computing capacity for map NameNodeRetryCache
2017-12-26 23:19:37,532 INFO util.GSet: VM type       = 64-bit
2017-12-26 23:19:37,533 INFO util.GSet: 0.029999999329447746% max memory 1.7 GB = 546.2 KB
2017-12-26 23:19:37,533 INFO util.GSet: capacity      = 2^16 = 65536 entries
2017-12-26 23:19:37,597 INFO namenode.FSImage: Allocated new BlockPoolId: BP-1437158662-192.168.18.160-1514301577580
2017-12-26 23:19:37,621 INFO common.Storage: Storage directory /opt/hadoop3/hdfs/name has been successfully formatted.
2017-12-26 23:19:37,644 INFO namenode.FSImageFormatProtobuf: Saving image file /opt/hadoop3/hdfs/name/current/fsimage.ckpt_00000000000000
2017-12-26 23:19:37,786 INFO namenode.FSImageFormatProtobuf: Image file /opt/hadoop3/hdfs/name/current/fsimage.ckpt_0000000000000000000 of
2017-12-26 23:19:37,811 INFO namenode.NNStorageRetentionManager: Going to retain 1 images with txid >= 0
2017-12-26 23:19:37,822 INFO namenode.NameNode: SHUTDOWN_MSG:
/************************************************************
SHUTDOWN_MSG: Shutting down NameNode at cdh1/192.168.18.160
************************************************************/
```

图 2-2 格式化 NameNode 成功结果

格式化时，需要注意 hadoop. tmp. dir 目录的权限问题，hadoop 普通用户应该有读写权限，可以将/opt/data 的所有者改为 hadoop。

```
[hadoop@bigdata1 hadoop-3.1.2]$ sudo chown -R hadoop:hadoop /apps/data
```

查看 NameNode 格式化后的目录，结果如图 2-3 所示。

```
[hadoop@bigdata1 hadoop-3.1.2]$ ll /apps/data/tmp/dfs/name/current
```

```
total 16
-rw-rw-r--. 1 hadoop hadoop 353 Jul 4 17:25 fsimage_0000000000000000000
-rw-rw-r--. 1 hadoop hadoop 62 Jul 4 17:25 fsimage_0000000000000000000.md5
-rw-rw-r--. 1 hadoop hadoop 2 Jul 4 17:25 seen_txid
-rw-rw-r--. 1 hadoop hadoop 202 Jul 4 17:25 VERSION
```

图 2-3 NameNode 格式化后目录结构

⑫ 启动 HDFS 和 YARN。

启动 NameNode：

```
[hadoop@bigdata001 hadoop-3.1.2]$ hadoop-daemon.sh start namenode
starting namenode, logging to/opt/modules/hadoop-3. 1. 2/logs/hadoop-hadoop-namenode-bigdata1.out
```

启动 DataNode：

```
[hadoop@bigdata001 hadoop-3.1.2]$ hadoop-daemon.sh start datanode
starting datanode, logging to /opt/modules/hadoop-3. 1. 2/logs/hadoop-hadoop-datanode-bigdata1.out
```

启动 SecondaryNameNode：

```
[hadoop@bigdata1 hadoop-3.1.2]$ hadoop-daemon.sh startsecondarynamenode
starting secondarynamenode, logging to/opt/modules/hadoop-3. 1. 2/logs/hadoop-hadoop-secondarynamenode-bigdata1.out
```

启动 ResourceManager：

```
[hadoop@bigdata001 hadoop-3.1.2]$ hadoop-daemon.sh start datanode
starting resourcemanager, logging to /opt/modules/hadoop-3. 1. 2/logs/hadoop-hadoop-namenode-bigdata1.out
```

启动 NodeManager：

```
[hadoop@bigdata001 hadoop-3.1.2]$ hadoop-daemon.sh start datanode
    starting nodemanager, logging to /opt/modules/hadoop-3.1.2/logs/hadoop-hadoop-namenode-bigdata1.out
```

⑬ 查看是否成功。

```
[hadoop@bigdata001 hadoop-3.1.2]$ jps
    3034 NameNode
    4439 NodeManager
    4197 ResourceManager
    4543 Jps
    3193 SecondaryNameNode
    3110 DataNode
```

通过网址“http://192.168.40.100”进行查看，如图 2-4 所示。

图 2-4　伪分布式 Hadoop 成功访问结果

⑭ 运行 MapReduce Job，此处运行自带的 WordCount 实例。

创建输入输出文件夹：

```
[hadoop@bigdata001 hadoop-3.1.2]$ hadoop hdfs dfs -mkdir -p /input
[hadoop@bigdata001 hadoop-3.1.2]$ hadoop hdfs dfs -mkdir -p /output
```

创建一个原始文件 word.txt：

```
[hadoop@bigdata001 hadoop-3.1.2]$ sudo vi /apps/data/word.txt
Hadoop mapreduce hive
Sqoop hadoop spark hbase
Storm hive
```

将 word.txt 文件上传到 HDFS 的/input 目录中：

```
[hadoop@bigdata1 hadoop-3.1.2]$ hdfs dfs -put /apps/data/word.txt /input
```

运行 WordCount MapReduce Job：

```
[hadoop@bigdata001 hadoop-3.1.2]$
hadoop jar share/hadoop/mapreduce/hadoop-mapreduce-examples-3.1.2.jar wordcount
/input /output/wordcountdemo
```

查看输出结果：

```
[hadoop@bigdata001 hadoop-3.1.2]$ hdfs dfs -ls /output/wordcountdemo
-rw-r--r-- 1 hadoop supergroup 0 2019-03-05 15:00 /output/wordcount/_SUCCESS
-rw-r--r-- 1 hadoop supergroup 60 2019-03-05 15:00 /output/wordcount/part-r-00000
```

查看文件内容：

```
[hadoop@bigdata001 hadoop-3.1.2]$ hdfs dfs -cat /output/wordcountdemo/part-
r-00000
hadoop  2
Hbase   1
Sqoop   1
Spark   1
Storm   1
Mapreduce  1
Hive  2
```

⑮ 开启历史服务。Hadoop 开启历史服务后可以在 Web 页面查看 YARN 上执行 Job 情况的详细信息，可以通过历史服务器查看已经运行完成的 MapReduce 作业记录，如用了多少 Map、用了多少 Reduce、作业提交时间、作业启动时间、作业完成时间等信息。

```
[hadoop@bigdata001 hadoop-3.1.2]$ mr-jobhistory-daemon.sh start historyerver
```

开启后，可以通过 Web 页面查看历史服务器：http://bigdata001:19888/。

2. 实验 2：完全分布模式

① 再开启 2 台虚拟机，主机名分别为 bigdata002、bigdata003；配置网络，修改网络参数。

```
bigdata002  192.168.40.101
bigdata003  192.168.40.102
```

② 配置 hosts(3 台机器)。

```
[hadoop@bigdata001 hadoop-3.1.2]$ sudo vi /etc/hosts
192.168.40.100 bigdata001
192.168.40.101 bigdata002
192.168.40.102 bigdata003
```

③ 配置 Windows 中的 SSH 客户端(通过 secureCRT 连接)。在本地 Windows 中的 SSH 客户端添加对 bigdata002、bigdata003 的 SSH 链接。

④ 服务器功能规划。

```
bigdata001          bigdata002          bigdata003
namenode            resourcemanager
datanode            datanode            datanode
nodemanager         nodemanager         nodemanager
historyserver                           secondarynamenode
```

⑤ 在第一台机器上安装新的 Hadoop，采用先在第一台机器上解压、配置 Hadoop，再分发到其他两台机器上的方式来安装集群。

```
[hadoop@bigdata001 modules]$ tar -zxvf hadoop-3.1.2.tar.gz
```

⑥ 配置 Hadoop JDK 路径，修改 hadoop-env.sh、mapred-env.sh、yarn-env.sh 文件中的 JDK 路径。

```
export JAVA_HOME = "/usr/local/java/jdk1.8.0_40"
```

⑦ 配置 core-site.xml。

```
[hadoop@bigdata001 hadoop-3.1.2]$ vi etc/hadoop/core-site.xml
<configuration>
    <property>
      <name>fs.defaultFS</name>
      <value>hdfs://bigdata001:8020</value>
    </property>
    <property>
      <name>hadoop.tmp.dir</name>
      <value>/apps/hadoop-3.1.2/data/tmp</value>
    </property>
</configuration>
```

fs.defaultFS 为 NameNode 的地址，hadoop.tmp.dir 为 Hadoop 临时目录的地址，默认情况下，NameNode 和 DataNode 的数据文件都会存在这个目录的对应子目录下，应保证此目录是存在的，如果不存在，应先创建。

⑧ 配置 hdfs-site.xml。

```
[hadoop@bigdata001 hadoop-3.1.2]$ vi etc/hadoop/hdfs-site.xml
<configuration>
    <property>
        <name>dfs.replication</name>
        <value>3</value>
    </property>
    <property>
        <name>dfs.namenode.secondary.http-address</name>
        <value>bigdata003:50090</value>
    </property>
</configuration>
```

dfs.namenode.secondary.http-address 指定 SecondaryNameNode 的 http 访问地址和端口号，因为在功能规划中已将 bigdata003 规划为 SecondaryNameNode 服务器，所以这里设置为 bigdata003:50090。

⑨ 配置 slaves。

```
[hadoop@bigdata001 hadoop-3.1.2]$ vi etc/hadoop/slaves
    bigdata001
    bigdata002
    bigdata003
```

slaves 文件用于指定 HDFS 上有哪些 DataNode 节点。

⑩ 配置 yarn-site.xml。

```
[hadoop@bigdata001 hadoop-3.1.2]$ vi etc/hadoop/yarn-site.xml

<configuration>
    <property>
          <name>yarn.nodemanager.aux-services</name>
          <value>mapreduce_shuffle</value>
    </property>
    <property>
          <name>yarn.resourcemanager.hostname</name>
          <value>bigdata002</value>
    </property>
    <property>
          <name>yarn.log-aggregation-enable</name>
          <value>true</value>
    </property>
    <property>
          <name>yarn.log-aggregation.retain-seconds</name>
          <value>106800</value>
    </property>
</configuration>
```

根据规划，yarn.resourcemanager.hostname 指定 ResourceManager 服务器指向 bigdata002，yarn.log-aggregation-enable 用于配置是否启用日志聚集功能，yarn.log-aggregation.retain-seconds 用于配置聚集的日志在 HDFS 上最多保存多长时间。

⑪ 配置 mapred-site.xml。

```
[hadoop@bigdata001 hadoop-3.1.2] $ vi etc/hadoop/mapred-site.xml

<configuration>
    <property>
        <name>mapreduce.framework.name</name>
        <value>yarn</value>
    </property>
    <property>
        <name>mapreduce.jobhistory.address</name>
        <value>bigdata001:10020</value>
    </property>
    <property>
        <name>mapreduce.jobhistory.webapp.address</name>
        <value>bigdata001:19888</value>
    </property>
</configuration>
```

mapreduce.framework.name 用于设置 MapReduce 任务运行在 YARN 上。mapreduce.jobhistory.address 用于设置 MapReduce 的历史服务器安装在 bigdata001 上，mapreduce.jobhistory.webapp.address 用于设置历史服务器的 Web 页面地址和端口号。

⑫ 设置 SSH 无密码登录。

在 bigdata001 上生成公钥：

```
[hadoop@bigdata001 hadoop-3.1.2] $ ssh-keygen -t rsa
```

按“Enter”键，都设置为默认值，然后在当前用户的 Home 目录下的.ssh 目录中会生成公钥文件(id_rsa.pub)和私钥文件(id_rsa)。

分发公钥：

```
[hadoop@bigdata001 hadoop-3.1.2] $ ssh-copy-id bigdata001
[hadoop@bigdata001 hadoop-3.1.2] $ ssh-copy-id bigdata002
[hadoop@bigdata001 hadoop-3.1.2] $ ssh-copy-id bigdata003
```

⑬ 设置 bigdata002、bigdata003 到其他机器的无密钥登录。同样地，在 bigdata002、bigdata003 上生成公钥和私钥后，将公钥分发到三台机器上。

⑭ 分发 Hadoop 文件。首先在其他两台机器上创建存放 Hadoop 的目录。

```
[hadoop@bigdata002 ~] $ mkdir /apps/modules
[hadoop@bigdata003 ~] $ mkdir /apps/modules
```

通过 Scp 分发。Hadoop 根目录下的 share/doc 目录中存放 Hadoop 的文档，文件相当大(约 1.6 GB)，建议在分发之前将这个目录删除，以节省硬盘空间并提高分发的速度。

```
[hadoop@bigdata001 hadoop-3.1.2]$ du -sh /apps/modules/hadoop-3.1.2/share/doc
1.6G    /apps/modules/hadoop-3.1.2/share/doc
[hadoop@bigdata001 hadoop-3.1.2]$ scp -r /apps/modules/hadoop-3.1.2/ bigdata002:/apps/modules
[hadoop@bigdata001 hadoop-3.1.2]$ scp -r /apps/modules/hadoop-3.1.2/ bigdata003:/apps/modules
```

⑮ 格式化 NameNode。

```
[hadoop@bigdata002 hadoop-3.1.2]$ /apps/modules/hadoop-3.1.2/bin/hdfs namenode -format
```

如果要重新格式化 NameNode，需要先将原来 NameNode 和 DataNode 下的文件全部删除，否则会报错，NameNode 和 DataNode 所在目录是在 core-site.xml 中的 hadoop.tmp.dir、dfs.namenode.name.dir、dfs.datanode.data.dir 属性配置的。

```
<property>
        <name>hadoop.tmp.dir</name>
        <value>/apps/data/tmp</value>
</property>
<property>
        <name>dfs.namenode.name.dir</name>
        <value>file://${hadoop.tmp.dir}/dfs/name</value>
</property>
<property>
        <name>dfs.datanode.data.dir</name>
        <value>file://${hadoop.tmp.dir}/dfs/data</value>
</property>
```

由于每次格式化时，默认会创建一个集群 ID，并写入 NameNode 和 DataNode 的 VERSION 文件中（VERSION 文件所在目录为 dfs/name/current 和 dfs/data/current），重新格式化时，默认会生成一个新的集群 ID，如果不删除原来的目录，则会导致 NameNode 的 VERSION 文件中是新的集群 ID，DataNode 中的是旧的集群 ID，不一致时会报错。

另一种方法是格式化时指定集群 ID 参数（指定为旧的集群 ID）。

⑯ 启动集群。

启动 HDFS：

```
[hadoop@bigdata001 hadoop-3.1.2]$ /apps/modules/hadoop-3.1.2/sbin/start-dfs.sh
```

启动 YARN：

```
[hadoop@bigdata001 hadoop-3.1.2]$ /apps/modules/hadoop-3.1.2/sbin/start-yarn.sh
```

在 bigdata002 上启动 ResourceManager：

```
[hadoop@bigdata002 hadoop-3.1.2]$ yarn-daemon.sh start resourcemanager
```

启动日志服务器：

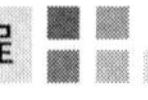

```
[hadoop@bigdata003 ~]$ /apps/modules/hadoop-3.1.2/sbin/mr-jobhistory-daemon.sh start historyserver
starting historyserver, logging to /apps/modules/hadoop-3.1.2/logs/mapred-hadoop-historyserver-bigda
ta0030.out
```

查看是否启动成功，若启动成功，则会出现图 2-2 所示信息：

```
[hadoop@bigdata003 ~]$ jps
3570 jps
3565 jobhistoryserver
3422 secondarynmenode
3212 datanode
3385 nodemanager
```

查看 HDFS Web 页面：http://bigdata-senior01.chybinmy.com:50070/。

查看 YARN Web 页面：http://bigdata002:8088/cluster。

⑰ 测试 Job。本实验用 Hadoop 自带的 WordCount 实例在本地模式下测试 MapReduce。

准备 MapReduce，输入 word.txt：

```
[hadoop@bigdata001 hadoop-3.1.2]$ cat /apps/data/word.txt
hadoop mapreduce hive
hbase spark storm hadoop
sqoop hadoop
spark hive
```

在 HDFS 中创建输入目录 input：

```
[hadoop@bigdata001 hadoop-3.1.2]$ hdfs dfs -mkdir /input
```

将 word.txt 上传到 HDFS：

```
[hadoop@bigdata001 hadoop-3.1.2]$ hdfs dfs /apps/data/word.txt /input
```

运行 Hadoop 自带的 mapreduce demo 实例：

```
[hadoop@bigdata001 hadoop-3.1.2]$ bin/yarn jar share/hadoop/mapreduce/hadoop-mapreduce-examples-3.1.2.jar wordcount /input/word.txt  /output
2019-03-16 23:43:58,173 WARN util.NativeCodeLoader: Unable to load native-hadoop library for your platform... using builtin-java classes where applicable
2019-03-16 23:43:59,210 INFO client.RMProxy: Connecting to ResourceManager at cdh1/192.168.40.100:8040
2019-03-16 23:43:59,817 INFO mapreduce.JobResourceUploader: Disabling Erasure Coding for path: /tmp/hadoop-yarn/staging/hadoop/.staging/job_1514302988215_0002
2019-03-16 23:44:01,017 INFO input.FileInputFormat: Total input files to process : 1
2019-03-16 23:44:01,198 INFO mapreduce.JobSubmitter: number of splits:1
```

```
2019-03-16 23:44:01,238 INFO Configuration.deprecation: yarn.resourcemanager.system-metrics-publisher.enabled is deprecated. Instead, use yarn.system-metrics-publisher.enabled
2019-03-16 23:44:01,387 INFO mapreduce.JobSubmitter: Submitting tokens for job: job_1514302988215_0002
2019-03-16 23:44:01,389 INFO mapreduce.JobSubmitter: Executing with tokens: []
2019-03-16 23:44:01,608 INFO conf.Configuration: resource-types.xml not found
2019-03-16 23:44:01,608 INFO resource.ResourceUtils: Unable to find 'resource-types.xml'.
2019-03-16 23:44:01,890 INFO impl.YarnClientImpl: Submitted application application_1514302988215_0002
2019-03-16 23:44:01,944 INFO mapreduce.Job: The url to track the job: http://bigdata001:8088/proxy/application_1514302988215_0002/
2019-03-16 23:44:01,945 INFO mapreduce.Job: Running job: job_1514302988215_0002
2019-03-16 23:44:11,098 INFO mapreduce.Job: Job job_1514302988215_0002 running in uber mode : false
2019-03-16 23:44:11,101 INFO mapreduce.Job:  map 0% reduce 0%
2019-03-16 23:44:19,223 INFO mapreduce.Job:  map 100% reduce 0%
2019-03-16 23:44:25,269 INFO mapreduce.Job:  map 100% reduce 100%
2019-03-16 23:44:25,290 INFO mapreduce.Job: Job job_1514302988215_0002 completed successfully
2019-03-16 23:44:25,468 INFO mapreduce.Job: Counters: 53
        File System Counters
                FILE: Number of bytes read=1963
                FILE: Number of bytes written=415199
                FILE: Number of read operations=0
                FILE: Number of large read operations=0
                FILE: Number of write operations=0
                HDFS: Number of bytes read=1758
                HDFS: Number of bytes written=1741
                HDFS: Number of read operations=8
                HDFS: Number of large read operations=0
                HDFS: Number of write operations=2
        Job Counters
                Launched map tasks=1
                Launched reduce tasks=1
                Data-local map tasks=1
                Total time spent by all maps in occupied slots (ms)=4962
```

```
                Total time spent by all reduces in occupied slots (ms) = 3408
                Total time spent by all map tasks (ms) = 4962
                Total time spent by all reduce tasks (ms) = 3408
                Total vcore-milliseconds taken by all map tasks = 4962
                Total vcore-milliseconds taken by all reduce tasks = 3408
                Total megabyte-milliseconds taken by all map tasks = 5081088
                Total megabyte-milliseconds taken by all reduce tasks = 3489792
        Map-Reduce Framework
                Map input records = 35
                Map output records = 55
                Map output bytes = 1885
                Map output materialized bytes = 1963
                Input split bytes = 93
                Combine input records = 55
                Combine output records = 54
                Reduce input groups = 54
                Reduce shuffle bytes = 1963
                Reduce input records = 54
                Reduce output records = 54
                Spilled Records = 108
                Shuffled Maps  = 1
                Failed Shuffles = 0
                Merged Map outputs = 1
                GC time elapsed (ms) = 100
                CPU time spent (ms) = 2130
                Physical memory (bytes) snapshot = 523571200
                Virtual memory (bytes) snapshot = 5573931008
                Total committed heap usage (bytes) = 443023360
                Peak Map Physical memory (bytes) = 302100480
                Peak Map Virtual memory (bytes) = 2781454336
                Peak Reduce Physical memory (bytes) = 221470720
                Peak Reduce Virtual memory (bytes) = 2792476672
        Shuffle Errors
                BAD_ID = 0
                CONNECTION = 0
                IO_ERROR = 0
                WRONG_LENGTH = 0
                WRONG_MAP = 0
```

```
                WRONG_REDUCE = 0
        File Input Format Counters
                Bytes Read = 1665
        File Output Format Counters
                Bytes Written = 1741
```

查看输出文件：

```
[hadoop@bigdata1 hadoop-3.1.2]$ hdfs dfs -ls /output/wordcountdemo
-rw-r--r-- 1 hadoop supergroup 0 2019-03-16 23:50 /output/_SUCCESS
-rw-r--r-- 1 hadoop supergroup 60 2019-03-16 23:50 /output/part-r-00000
```

2.7 参考答案

实验作业的答案见本章实验内容部分。

第3章

大数据处理:分布式处理框架 Spark 实验教程

3.1 Spark 安装

1. 实验目的

了解 Spark 的不同部署模式,学会部署并启动 Spark 集群,能够配置 Spark 集群使用 HDFS。

2. 实验要求

① 构建出完整的 Spark 集群:在 master 上部署主服务 Master;在 slave1、slave2、slave3 上部署从服务 Worker。

② 完成部署之后,要求能成功访问 Spark UI 并能成功执行 Spark 样例程序。

3. 预备知识

拓展阅读

Spark 介绍

Apache Spark 基于开源的集群运算框架。经过 2013—2014 年的高速发展,Spark 目前已经成为大数据计算领域最热门的技术之一。Spark 的核心技术弹性分布式数据集(RDD,Resilient Distributed Datasets)提供了比 Hadoop 更加丰富的 MapReduce 模型,拥有 Hadoop MapReduce 所具有的所有优点,但不同于 Hadoop MapReduce 的是,Spark 中 Job 的中间输出和结果可以保存在内存中,从而可以基于内存快速地对数据集进行多次迭代,以此来支持复杂的机器学习、图计算和准实时流处理等,效率更高,速度更快。自从 Spark 将其代码部署到 GitHub 上之后,截至 2018 年 11 月共有 23 093 次提交,19 个分支,82 次发布,1 296 位代码贡献者,可以发现 Spark 开源社区的活跃度相当高,Spark 是目前最受欢迎的集群运算框架之一。Spark 官方网站的网址是 http://spark.apache.org/。截至 2018 年 11 月,Spark 的最新版本是 Spark 2.3.2。

4. 实验环境

需要 4 台安装 Linux 系统的物理机或虚拟机,要求 4 个 Linux 之间能 SSH 免密访问,安装 Java 环境,配置 hosts 文件。本书使用 4 台 VirtualBox 虚拟机,配置如表 3-1 所示。

表 3-1

IP 地址	节点名称	系统	内存	JDK 版本	Spark 版本	Hadoop 版本
192.168.56.100	master	CentOS 7 x64	4 GB	1.8	2.3.2	2.7.7
192.168.56.101	slave1	CentOS 7 x64	2 GB	1.8	2.3.2	2.7.7
192.168.56.102	slave2	CentOS 7 x64	2 GB	1.8	2.3.2	2.7.7
192.168.56.103	slave3	CentOS 7 x64	2 GB	1.8	2.3.2	2.7.7

5. 实验步骤

（1）下载 Spark

① 进入官方网站 http://spark.apache.org[13]，如图 3-1 所示，选择“Download”。

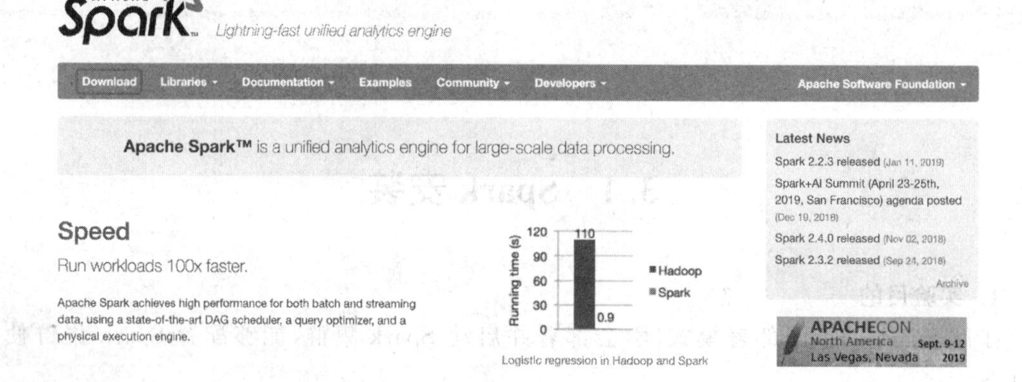

图 3-1 Spark 官方首页

② “Choose a Spark release”处选择“2.3.2”版本，“Choose a package type”处选择“Pre-built for Apache Hadoop 2.7 and later”(针对 Hadoop 2.7 及以后版本的二进制版本)，如图 3-2 所示。

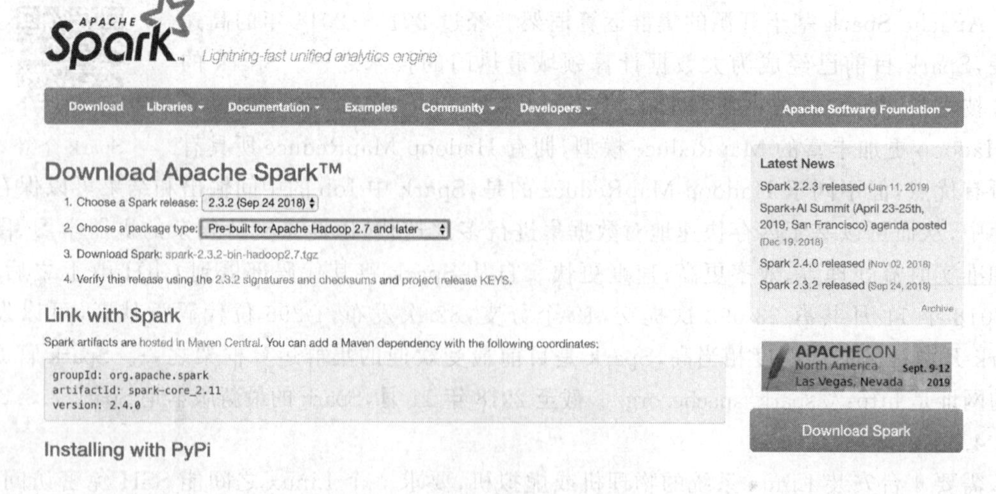

图 3-2 Spark 官方下载页

③ 点击下载“spark-2.3.2-bin-hadoop2.7.tgz”，并将下载的文件传入 master 虚拟机的

/usr/local/src 目录下，然后进入/usr/local/src 目录，解压并修改文件名为 spark。

解压命令：

```
tar -xzvf spark-2.3.2-bin-hadoop2.7.tgz
```

修改文件名命令：

```
mv spark-2.3.2-bin-hadoop2.7 spark
```

（2）配置 Spark 环境

在 master 上执行以下操作（Spark 独立模式）。

① 进入/usr/local/src/spark/conf 目录。

```
cd /usr/local/src/spark/conf
```

② 复制 slaves.template 文件为 slaves。

```
cp slaves.template slaves
```

③ 将 slave 节点的名称加入 slaves 文件，表示当前的 Spark 集群共有 3 个 slave 节点，节点的名称分别是 slave1、slave 2、slave3（原文件中可能有 localhost，需要删除，否则会在 master 上启动 Worker 进程）。

编辑命令：

```
vim slaves
```

加入内容：

```
slave1
slave2
slave3
```

④ 在 spark-config 中加入 JAVA_HOME。

编辑 spark-config 命令：

```
vim /usr/local/src/spark/sbin/spark-config.sh
```

加入内容：

```
export JAVA_HOME = /usr/local/src/java
```

⑤ 将配置好的 Spark 复制到 slave1、slave2、slave3 的/usr/local/src 目录下。

复制命令：

```
scp -r /usr/local/src/spark root@slave1: /usr/local/src
scp -r /usr/local/src/spark root@slave2: /usr/local/src
scp -r /usr/local/src/spark root@slave3: /usr/local/src
```

注意：此时已经可以独立启动 Spark 集群，启动方式见下述内容。

（3）配置 HDFS

① 将 Spark 环境变量模板复制成环境变量文件[14]，复制命令：

```
cp /usr/local/src/spark/conf/spark-env.sh.template   /usr/local/src/spark/conf/spark-env.sh
```

② 修改 spark-env.sh 文件，在文件末尾添加 Hadoop 配置文件的路径，添加内容：

```
export HADOOP_CONF_DIR = /usr/local/src/hadoop/etc/hadoop
```

(4) 启动 Spark 集群

在 master 上执行"/usr/local/src/spark/sbin/start-all. sh",启动 Spark 集群(若要使用 HDFS,则需要手动启动 Hadoop)。

(5) 提交 Spark 任务

使用 Shell 命令向 Spark 集群提交 Spark-App。

① 创建 HDFS 目录。

```
hadoop fs -mkdir -p /user/spark/in
```

② 创建输入文件 test1_in. txt,文件内容:

```
hello world
hello my friend
hello friend
```

③ 上传 test1_in. txt 文件到 HDFS 目录/user/spark/in 下,文件上传命令:

```
hadoop fs -put ./test1_in.txt /user/spark/in
```

④ 提交 WordCount 示例代码,执行命令:

```
/usr/local/src/spark/bin/spark-submit --master spark://master:7077 \
--class org.apache.spark.examples.JavaWordCount\
/usr/local/src/spark/examples/jars/spark-examples_2.11-2.3.2.jar \
hdfs://master:9000/user/spark/in/test1_in.txt
```

注意:若命令需要换行,则应在后面加上"\"。

(6) 实验结果

① 查看进程。

```
jps
```

master 中的进程如下:

```
9155 ResourceManager
8807 NameNode
11687 Master
11799 Jps
9001 SecondaryNameNode
```

slave 中的进程如下:

```
4385 NodeManager
4275 DataNode
5731 Jps
5610 Worker
```

若 master 节点的进程中有 Master,slave 节点的进程中有 Worker,则说明 Spark 集群进程成功启动。

② 查看 WebUI。若 Spark 集群成功启动,则应当能成功访问 Spark 的 WebUI 地址 http://master:8080(虚拟机需要关闭防火墙才能成功访问),页面如图 3-3 所示。

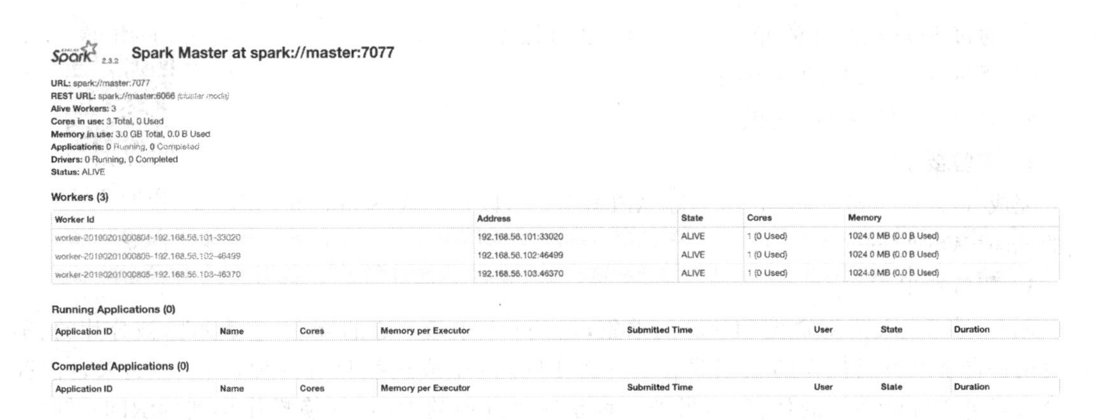

图 3-3 Spark-Web 页面

③ SparkWordCount 查询执行结果。Shell 输出结果如下所示：

```
2019-02-01 00：21：28 INFO   DAGScheduler：54 - Job 0 finished： collect at
JavaWordCount.java:53, took 3.916964 s
  hello：3
  my：1
  friend：2
  world：1
2019-02-01 00:21:28 INFO  AbstractConnector:318 - Stopped Spark@1a15b789{HTTP/
1.1,[http/1.1]}{0.0.0.0:4040}
2019-02-01 00:21:28 INFO  SparkUI:54 - Stopped Spark web UI at http://slave1:4040
```

WebUI 中的详细信息如图 3-4 所示。

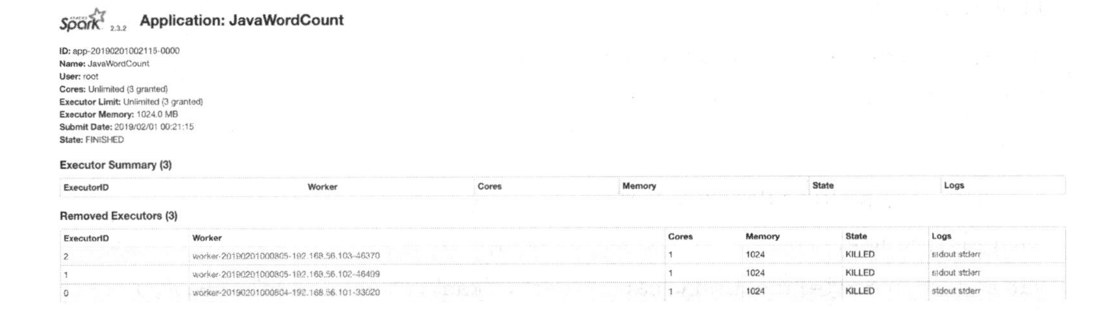

图 3-4 WebUI 中显示的任务信息

3.2 Spark-shell

通过 Spark-shell 执行 RDD 操作，示例程序使用 WordCount。

1. 实验目的

① 了解 Spark-shell 的操作方式。

② 通过 Spark-shell 的操作方式理解 RDD 操作。

③ 通过 RDD 操作理解 RDD 的原理。

④ 对 Scala 有一定的认识。

拓展阅读

RDD 介绍

2. 实验要求

完成 max、first、distinct、foreach 等 API 操作[15]，以及完成 Spark-shell 形式的 SparkWordCount。

3. 预备知识

RDD 是 Spark 中对数据和计算的抽象，是 Spark 中最核心的概念，它表示已被分片的、不可变的并能够被并行操作的数据集合。一个 RDD 的生成途径只有两种，第一种是来自内存集合或者外部存储系统的数据集，另外一种是通过其他的 RDD 转换操作得到的 RDD，例如，RDD 可以通过 map、filter、join 等操作转换为另一个 RDD。在 Spark 中，对于 RDD 的操作一般可以分为两种：转换操作（transformation）和行动操作（action）。

① 转换操作：将 RDD 通过一定的操作变为另一个 RDD，例如，将 file 这个 RDD 通过一个 filter 操作变换成 filterRDD，所以 filter 是一个转换操作。

② 行动操作：由于 Spark 是惰性计算的，因此对任何 RDD 进行行动操作，都会触发 Spark 作业的运行，从而产生最终的结果。例如，对 filterRDD 进行的 count 操作就是一个行动操作，即能使 RDD 产生结果的操作是行动操作。

Spark 数据处理程序会通过创建 RDD（输入）、转换操作、行动操作（输出）来完成一个作业。

4. 实验环境

需要 4 台安装 Linux 系统的物理机或虚拟机，本书使用 4 台 VirtualBox 虚拟机，配置如表 3-1 所示。

5. 实验步骤

进入 Spark 目录：

```
cd /usr/local/src/spark/
```

启动 Spark-shell：

```
bin/spark-shell --master spark://master:7077
```

退出 Spark-shell：

```
quit
```

（1）distinct：去除 RDD 内的重复数据

shell 输入 Scala 语句：

```
var a = sc.parallelize(List("car", "car", "man", "man", "wen", "boy"))
a.distinct.collect
```

shell 执行输出结果如下：

```
scala> var a = sc.parallelize(List("car", "car", "man", "man", "wen", "boy"));
a: org.apache.spark.rdd.RDD[String] = ParallelCollectionRDD[12] at parallelize
                                       at <console>:24


scala> a.distinct.collect
res7: Array[String] = Array(boy, wen, car, man)
```

(2) foreach:遍历 RDD 内的数据

shell 输入 Scala 语句：

```
var b = sc.parallelize(List("car", "she", "man", "he", "wen", "boy" ,"cat" ,"you"));
b.collect.foreach(x => println(x + " is my friend "));
```

shell 执行输出结果如下：

```
scala> var b = sc.parallelize(List("car", "she", "man", "he", "wen", "boy" ,
                                   "cat" ,"you"));
b: org.apache.spark.rdd.RDD[String] = ParallelCollectionRDD[1] at parallelize
                                      at <console>:24

scala> b.collect.foreach(x => println(x + " is my friend "));
car is my friend
she is my friend
man is my friend
he is my friend
wen is my friend
boy is my friend
cat is my friend
you is my friend
```

(3) first:取得 RDD 中的第一个元素

shell 输入 Scala 语句：

```
var c = sc.parallelize(List("car", "car", "man", "man", "wen", "boy"))
c.first
```

shell 执行输出结果如下：

```
scala> var c = sc.parallelize(List("car", "car", "man", "man", "wen", "boy"))
c: org.apache.spark.rdd.RDD[String] = ParallelCollectionRDD[2] at parallelize
                                      at <console>:24

scala> c.first
res4: String = car
```

(4) max:取得 RDD 中的最大元素

shell 输入 Scala 语句：

```
var d = sc.parallelize(10 to 30)
d.max
```

shell 执行输出结果如下：

```
scala> var d = sc.parallelize(10 to 30)
d: org.apache.spark.rdd.RDD[Int] = ParallelCollectionRDD[3] at parallelize at
                                   <console>:24

scala> d.max
res5: Int = 30
```

(5) intersection:返回两个 RDD 重叠元素

shell 输入 Scala 语句：

```
var f = sc.parallelize(1 to 20);
var g = sc.parallelize(10 to 30);
var h = f.intersection (g);
h.collect.sorted;
```

shell 执行输出结果如下：

```
scala> var f = sc.parallelize(1 to 20)
f: org.apache.spark.rdd.RDD[Int] = ParallelCollectionRDD[20] at parallelize at
                                   <console>:24

scala> var g = sc.parallelize(10 to 30)
g: org.apache.spark.rdd.RDD[Int] = ParallelCollectionRDD[21] at parallelize at
                                   <console>:24

scala> var h = f.intersection (g)
h: org.apache.spark.rdd.RDD[Int] = MapPartitionsRDD[27] at intersection at
                                   <console>:27

scala> h.collect.sorted
res13: Array[Int] = Array(10, 11, 12, 13, 14, 15, 16, 17, 18, 19, 20)
```

(6) 编写 WordCount

shell 输入 Scala 语句：

```
var file = sc.textFile("hdfs://master:9000/user/spark/in/test1_in.txt");
var count = file.flatMap(line => line.split(" ")).map(word =>(word,1)).
         reduceByKey(_+_);
count.collect;
```

shell 执行输出结果如下：

```
scala> var file = sc.textFile("hdfs://master:9000/user/spark/in/test1_in.txt")
file: org.apache.spark.rdd.RDD[String] = hdfs://master:9000/user/spark/in/test1_in.txt MapPartitionsRDD[29] at textFile at <console>:24

scala> var count = file.flatMap(line => line.split(" ")).map(word =>(word,1)).
                reduceByKey(_ + _)
count: org.apache.spark.rdd.RDD[(String, Int)] = ShuffledRDD[32] at reduceByKey at <
                                                    console>:25

scala> count.collect
res15: Array[(String, Int)] = Array((hello,3), (my,1), (friend,2), (world,1))
```

(7) 实验结果

本次实验结果均已给出。

3.3 Spark Scala

1. 实验目的

学会使用 IntelliJ IDEA 集成 Scala 开发环境，掌握 Spark 远程提交方式。[16]

2. 实验要求

使用 IntelliJ IDEA 创建 Scala 环境的 Spark 项目，完成倒排索引(Inverted Index)功能的代码编写。将程序代码打包为 jar，提交到 Spark 集群执行。

倒排索引源于实际应用中需要依据内容查找索引，如搜索引擎中根据搜索的内容查找需要的网址，则需要建立倒排索引。[17]

例如，输入内容：

```
id1 the spark
id2 the hdfs
id3 hbase
id4 the spark
id5 hive
id6 spark
id7 hdfs and spark
id8 the yarn
```

输出结果：

```
hive id5
yarn id8
spark id6 id7 id9 id1 id4
and id7
```

```
hdfs id7 id2
the id8 id1 id2 id4
hbase id3
```

3. 预备知识

拓展阅读
Scala 基本语法

Scala 是一门类 Java 的编程语言，它结合了面向对象编程和函数式编程。Scala 是纯面向对象的，每个值都是一个对象，对象的类型和行为由类定义，不同的类可以通过混入（mixin）的方式组合在一起。Scala 的设计目的是要和两种主流面向对象编程语言 Java 和 C#实现无缝互操作，这两种主流语言都非纯面向对象。

Scala 也是一门函数式编程语言，每个函数都是一个值，原生支持嵌套函数定义和高阶函数。Scala 也支持一种通用形式的模式匹配，模式匹配用于操作代数式类型，在很多函数式语言中都有实现。

Scala 被设计用于和 Java 无缝互操作。Scala 类可以调用 Java 方法、创建 Java 对象、继承 Java 类和实现 Java 接口，这些都不需要额外的接口定义或者胶合代码。Scala 始于 2001 年，由洛桑联邦理工学院（EPFL）的编程方法实验室研发，2003 年 11 月发布 1.0 版本。

4. 实验环境

客户端环境：安装 IntelliJ IDEA 和 JDK 1.8。

服务端环境：需要 4 台安装 Linux 系统的物理机或虚拟机，本书使用 4 台 VirtualBox 虚拟机，配置如表 3-1 所示。

5. 实验步骤

（1）安装 Scala 插件

① 打开 IntelliJ IDEA，选择“Configure”，如图 3-5 所示，再选择“Plugins”，如图 3-6 所示。

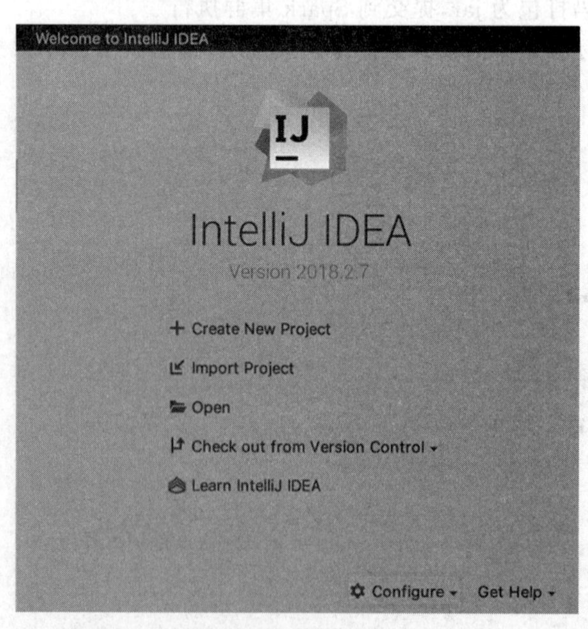

图 3-5　IntelliJ IDEA 界面（一）

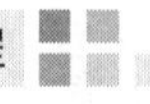

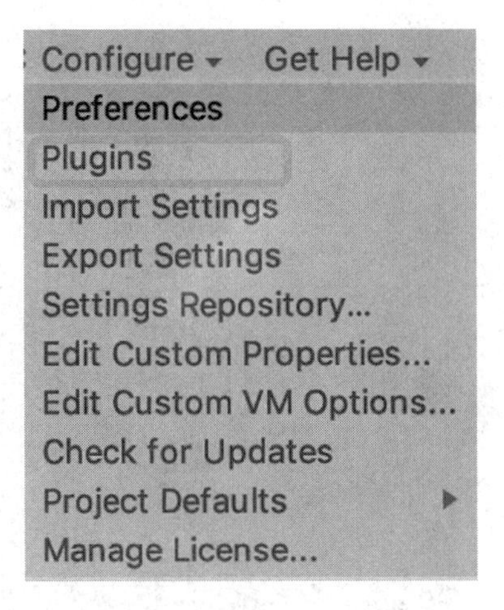

图 3-6 IntelliJ IDEA 界面(二)

② 进入 Plugins 之后，选择“Browse repositories...”，如图 3-7 所示。

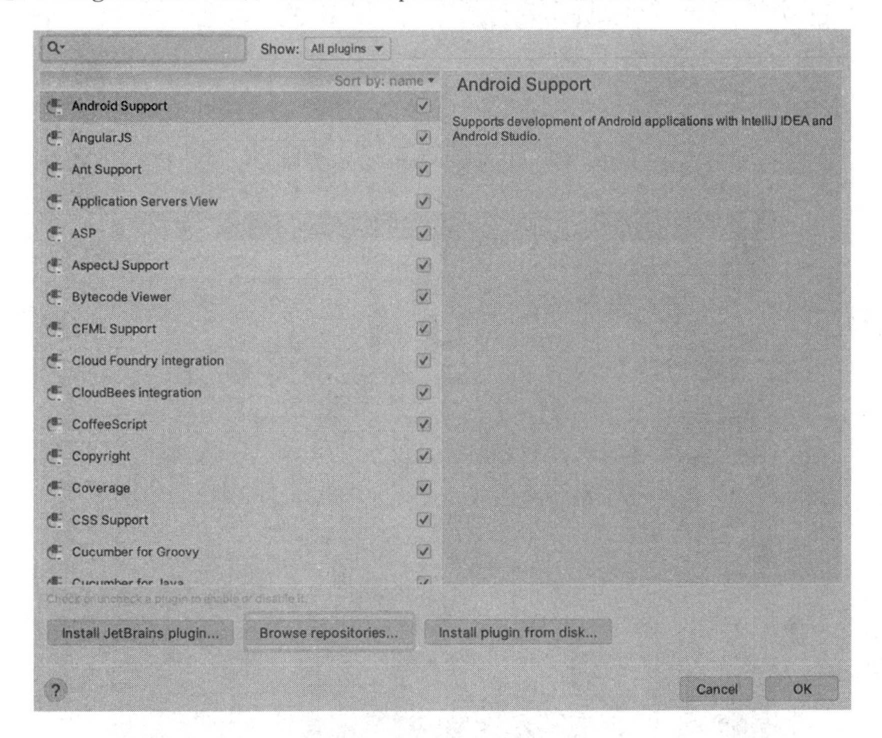

图 3-7 Plugins 界面

③ 在搜索框内输入 scala，选择列表中下载量最多的一项，单击“Install”，如图 3-8 所示。

④ Scala 插件下载完成后，重启 IntelliJ IDEA。

(2) 建立 Scala 项目

选用 Maven 作为项目的包管理工具，在 IntelliJ IDEA 中新建一个 Maven+Spark+Scala 项目。

① 打开 IntelliJ IDEA，选择“Create New Project”，如图 3-9 所示。

② 在新建项目界面，选择“Maven”，JDK 版本选择“1.8”，单击“Next”，如图 3-10 所示。

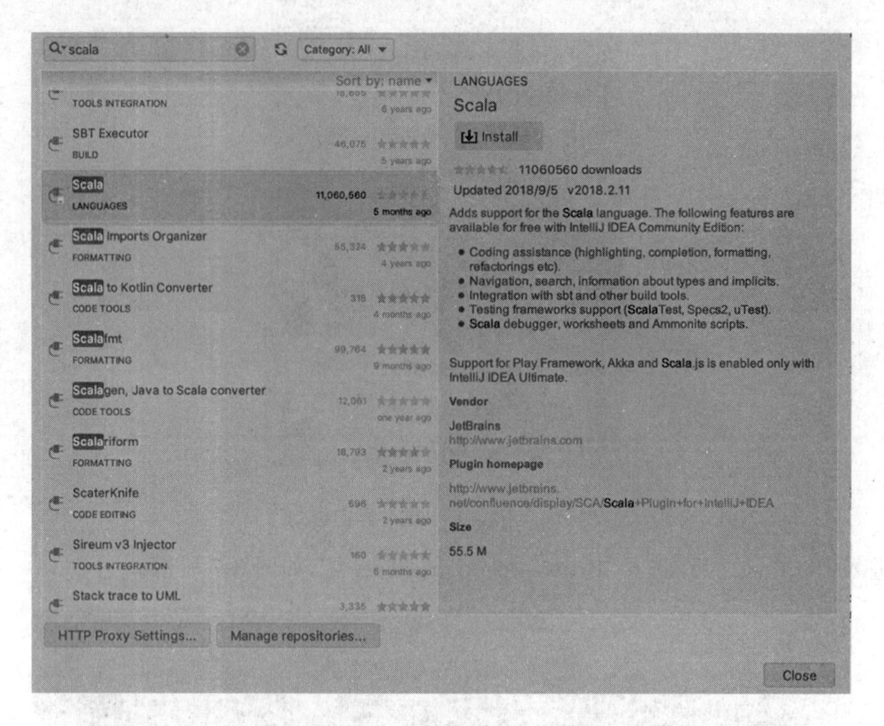

图 3-8 Browse repositories 界面

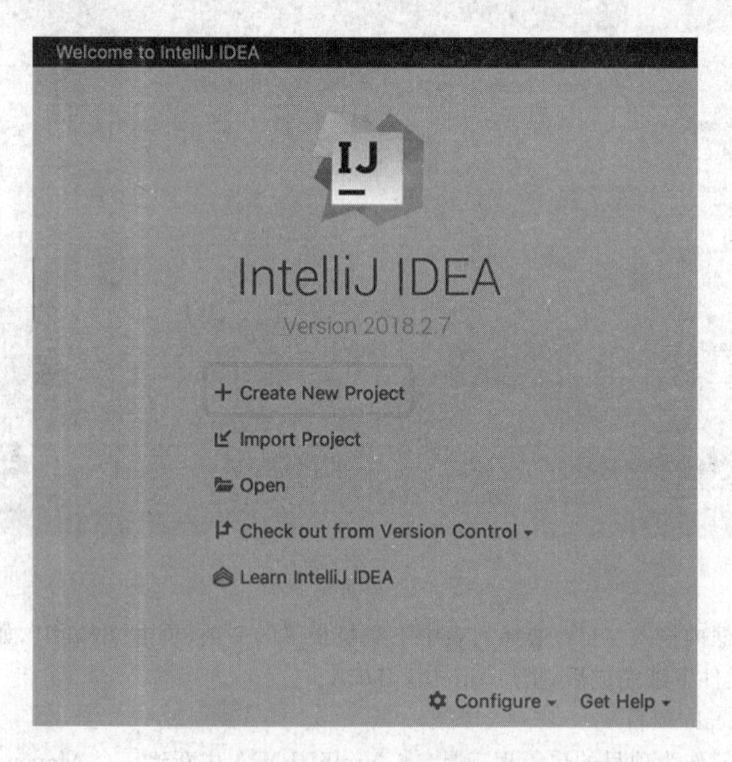

图 3-9 IntelliJ IDEA 界面(三)

③ 输入组织名称和项目名称,单击“Next”,如图 3-11 所示,再默认单击“Finish”,如图 3-12 所示。

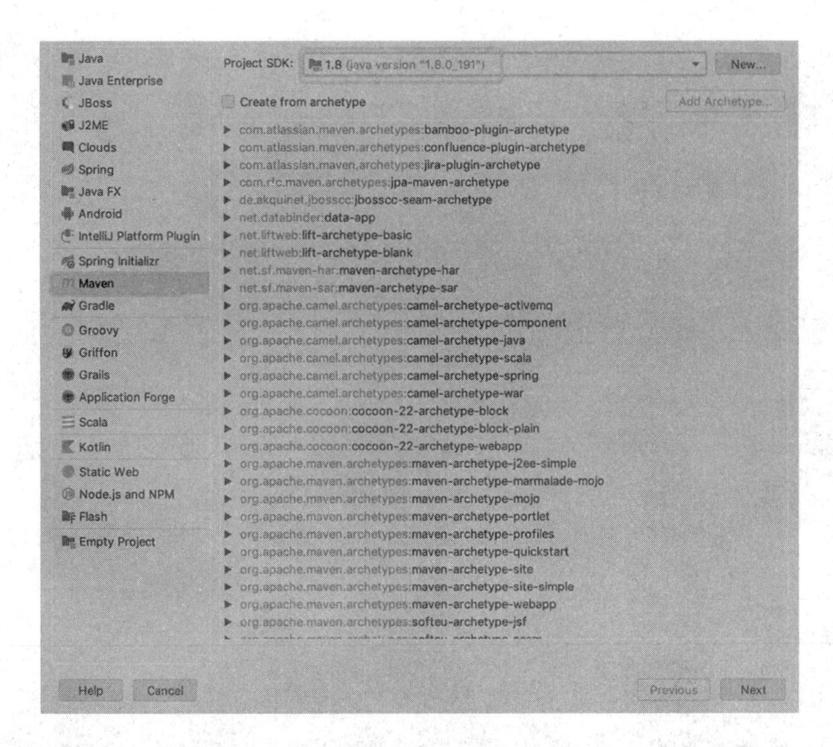

图 3-10 Create New Project 界面(一)

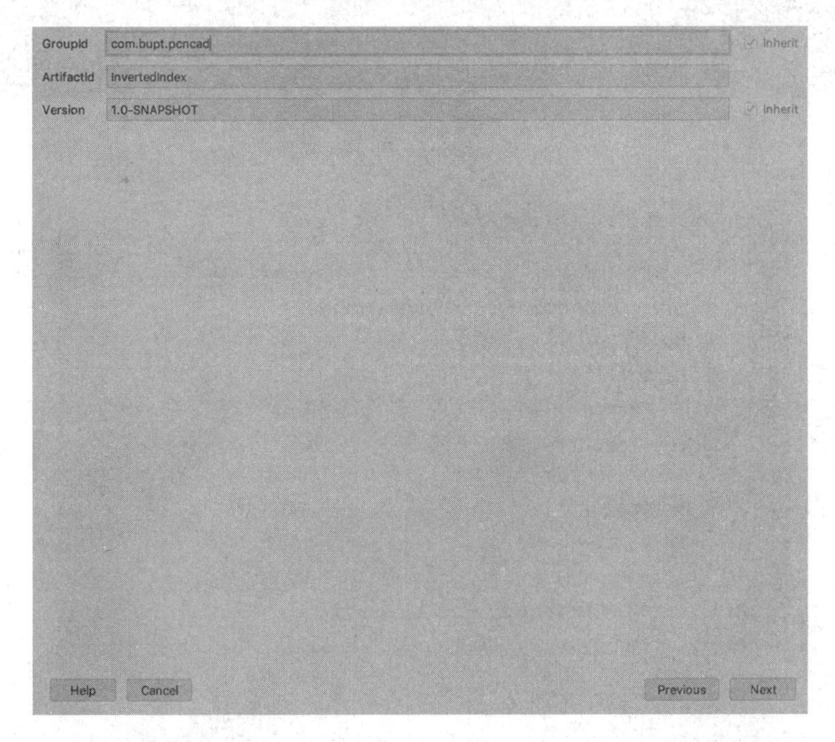

图 3-11 Create New Project 界面(二)

④ 下面开始添加 Maven 项目的包依赖，修改 pom. xml 文件，并将 src/main/目录下的 java 文件夹重命名为 scala(如果没有则手动新建 scala 文件夹，并选中 scala 文件夹，右击选择“Mark Directory as→Sources Root”)，如图 3-13 所示。

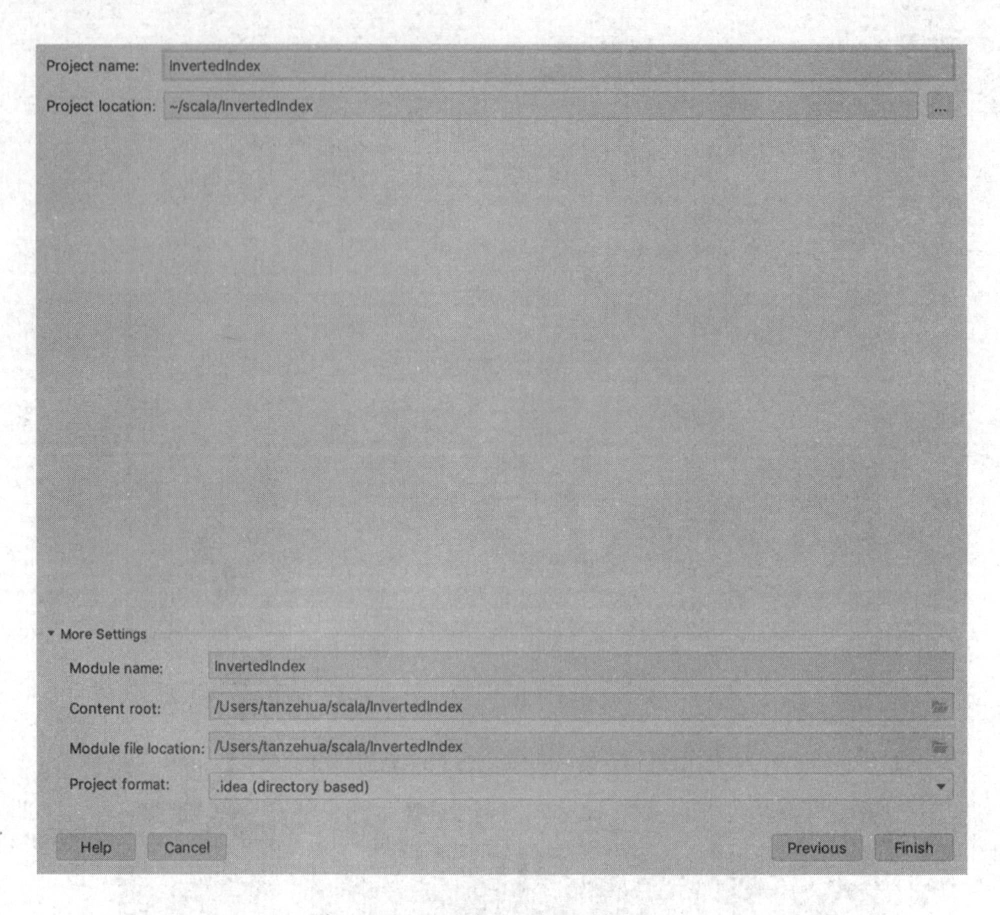

图 3-12　Create New Project 界面(三)

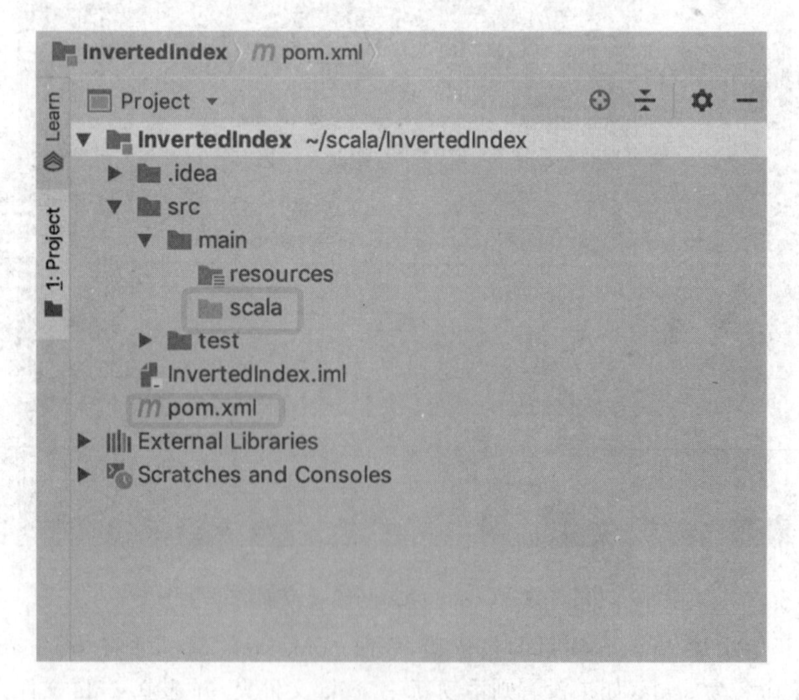

图 3-13　目录结构

pom.xml 文件内容如下所示。

```
<?xml version="1.0" encoding="UTF-8"?>
<project xmlns="http://maven.apache.org/POM/4.0.0"
        xmlns:xsi="http://www.w3.org/2001/XMLSchema-instance"
        xsi:schemaLocation="http://maven.apache.org/POM/4.0.0
                            http://maven.apache.org/xsd/maven-4.0.0.xsd">
    <modelVersion>4.0.0</modelVersion>

    <groupId>com.bupt.pcncad</groupId>
    <artifactId>InvertedIndex</artifactId>
    <version>1.0-SNAPSHOT</version>
    <properties>
        <spark.version>2.3.2</spark.version>
        <scala.version>2.11</scala.version>
    </properties>
    <dependencies>
      <dependency>
        <groupId>org.apache.spark</groupId>
        <artifactId>spark-core_${scala.version}</artifactId>
        <version>${spark.version}</version>
      </dependency>
    </dependencies>
    <build>
      <plugins>
        <plugin>
          <groupId>org.scala-tools</groupId>
          <artifactId>maven-scala-plugin</artifactId>
          <version>2.15.2</version>
          <executions>
            <execution>
                <goals>
                  <goal>compile</goal>
                  <goal>testCompile</goal>
                </goals>
            </execution>
          </executions>
        </plugin>
      </plugins>
    </build>
</project>
```

⑤ 在./src/main/scala 目录下新建 Scala Object 文件，右击 scala 目录选择“new→scala class”，类型选择“Object”，单击“OK”，如图 3-14 所示。

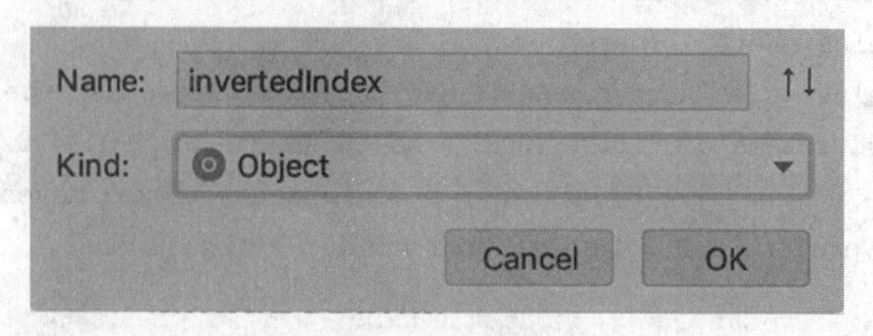

图 3-14 Create New Scala Class 界面

⑥ 选择“File→Project Structure→Artifacts”，单击“+”，选择“JAR→From modules with dependencies”，如图 3-15 所示，再选择“copy to the output directory and link via manifest”，单击“OK”，如图 3-16 所示。

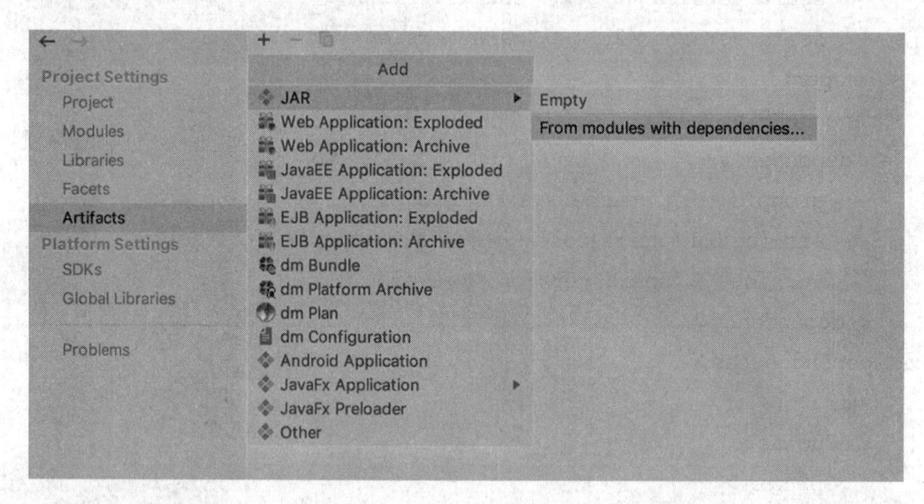

图 3-15 Artifacts 界面(一)

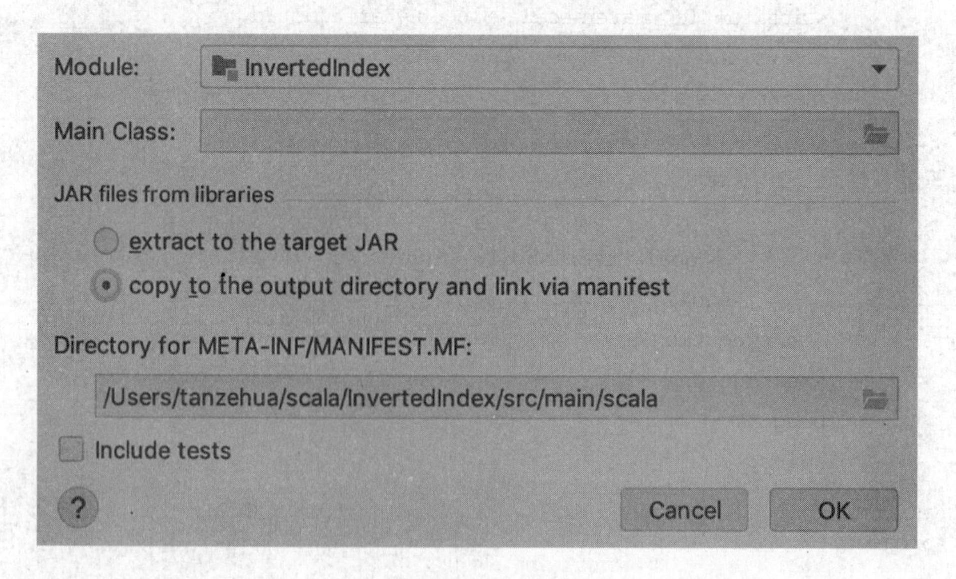

图 3-16 Artifacts 界面(二)

⑦ 上传输入文件至 HDFS：

```
hadoop fs -put -f  /opt/test2_in.txt /user/spark/in/
```

文件内容：

```
id1 the spark
id2 the hdfs
id3 hbase
id4 the spark
id5 hive
id6 spark
id7 hdfs and spark
id8 the yarn
```

(3) 编写 Inverted Index 程序[18]

① 编写代码，如下所示。

```
import org.apache.spark.{SparkConf, SparkContext}

object invertedIndex {
  def main(args: Array[String]){
    System.setProperty("HADOOP_USER_NAME", "root")
    val conf = new SparkConf().setAppName("invertedIndex")
      .setMaster("spark://192.168.56.100:7077")
      .setJars(List("./out/artifacts/InvertedIndex_jar/InvertedIndex.jar"))
    val sc = new SparkContext(conf)
    // 读取文件
    val input = sc.textFile("hdfs://192.168.56.100:9000/user/spark/in/test2_
                            in.txt")
    // 截取 ID 为 key,每行的内容为 value,组成(key,value)的形式
    val lines = input.map(line => (line.split(" ")(0), line))
    // 将每行的单词拆分出来和该行的 ID 组合
    val word = lines.flatMapValues(line => line.substring(line.indexOf(" ") +
                                   1).split(" "))
    // 翻转 ID 和 word
    val invertedWord = word.map(term => (term._2,term._1))
    // 将相同 word 的 ID 组合到一起
    val result = invertedWord.reduceByKey((x,y) => x + " " + y)
    // 输出结果
    result.collect().foreach(x => println(x))
  }
}
```

注意：“.setJars(List("./out/artifacts/InvertedIndex_jar/InvertedIndex.jar"))”中的路

径是“Artifacts”中设置的 Jar 生成路径。

(4) 执行 Inverted Index 程序

① 右击新建的 Object 文件，选择“Create‘InvertedIndex’”，单击下方的“+”，如图 3-17 所示。

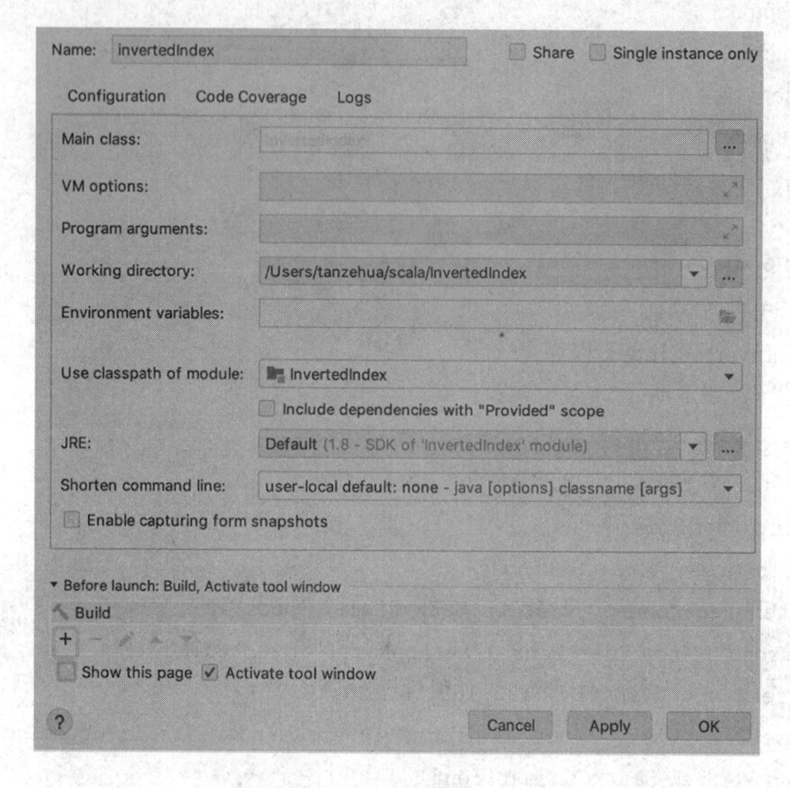

图 3-17　Create InvertedIndex 界面(一)

② 单击“+”后，选择“Build Artifacts”，勾选“InvertedIndex:jar”，单击“OK”，如图 3-18 和图 3-19 所示。

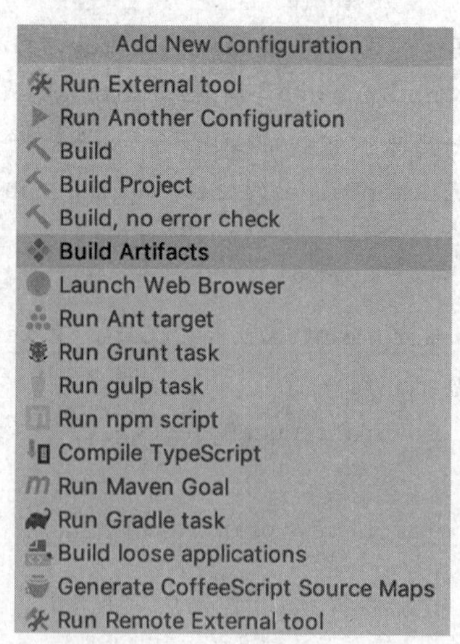

图 3-18　Create InvertedIndex 界面(二)

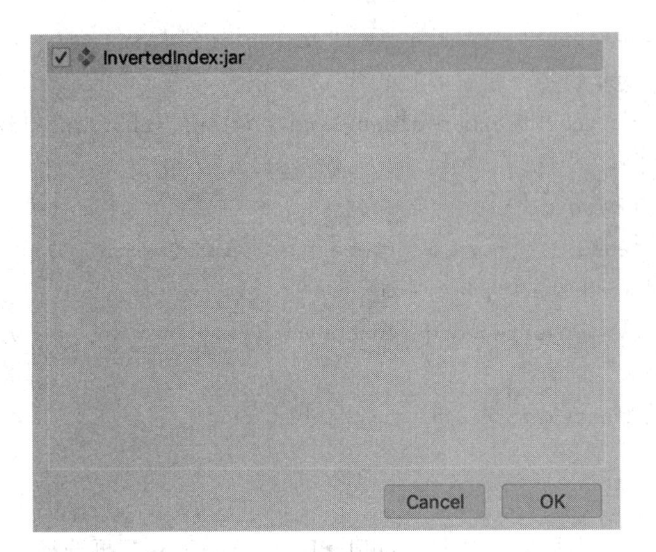

图 3-19 Create InvertedIndex 界面(三)

③ 单击 IDEA 右上角的运行按钮，执行代码。

④ 执行结果如图 3-20 所示。

```
19/02/11 16:33:53 INFO TaskSchedulerImpl: Removed TaskSet 1.0, whose tasks have all completed, from pool
19/02/11 16:33:53 INFO DAGScheduler: ResultStage 1 (collect at invertedIndex.scala:16) finished in 0.168 s
19/02/11 16:33:53 INFO DAGScheduler: Job 0 finished: collect at invertedIndex.scala:16, took 4.957010 s
(hive,id5)
(yarn,id8)
(spark,id1 id4 id6 id7 id9)
(hdfs,id2 id7)
(and,id7)
(the,id1 id2 id4 id8)
(hbase,id3)
19/02/11 16:33:53 INFO SparkContext: Invoking stop() from shutdown hook
19/02/11 16:33:53 INFO SparkUI: Stopped Spark web UI at http://192.168.3.104:4040
19/02/11 16:33:53 INFO StandaloneSchedulerBackend: Shutting down all executors
```

图 3-20 远程提交结果

(5) 打包成 jar 执行

由于将代码提交给其他用户运行时，难以保证大家的开发环境一致，因此往往将代码打包成 jar，然后上传至服务器运行。

① 修改部分代码，将“.setJars(List(\"./out/artifacts/InvertedIndex_jar/InvertedIndex.jar\"))”删除，如下所示。

```
import org.apache.spark.{SparkConf, SparkContext}
object invertedIndex {
  def main(args: Array[String]){
    System.setProperty("HADOOP_USER_NAME", "root")
    val conf = new SparkConf().setAppName("invertedIndex")
      .setMaster("spark://192.168.56.100:7077")
    val sc = new SparkContext(conf)
    // 读取文件
    val input = sc.textFile("hdfs://192.168.56.100:9000/user/spark/in/test2_in.txt")
    // 截取 ID 为 key,每行的内容为 value,组成(key,value)的形式
```

```
    val lines = input.map(line => (line.split(" ")(0), line))
    // 将每行的单词拆分出来和该行的 ID 组合
    val word = lines.flatMapValues(line => line.substring(line.indexOf(" ") +
                                    1).split(" "))
    // 翻转 ID 和 word
    val invertedWord = word.map(term => (term._2,term._1))
    // 将相同 word 的 ID 组合到一起
    val result = invertedWord.reduceByKey((x,y) => x + " " + y)
    // 输出结果
    result.collect().foreach(x => println(x))
  }
}
```

② 修改代码后，新编译项目，选择“Build→Build Artifacts”，再选择“InvertedIndex:jar→Build”，如图 3-21 所示。

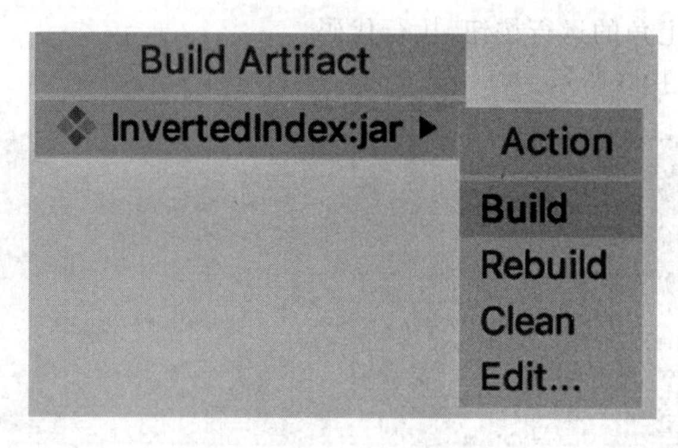

图 3-21　Build 界面

③ 将重新编译的 jar 包上传至 master 的/opt 目录下。

④ 修改连接 master 命令 shell 界面，进入 spark 目录下执行以下命令。

```
./bin/spark-submit \
--master spark://master:7077\
--class invertedIndex /opt/InvertedIndex.jar
```

⑤ 运行结果如下：

```
  2019-02-11 17:24:28 INFO   DAGScheduler:54 - Job 0 finished: collect at
invertedIndex.scala:14, took 4.339240 s
  (hive,id5)
  (yarn,id8)
  (spark,id6 id7 id9 id1 id4)
  (and,id7)
  (hdfs,id7 id2)
  (the,id8 id1 id2 id4)
```

```
(hbase,id3)
2019-02-11 17:24:28 INFO  SparkContext:54 - Invoking stop() from shutdown hook
2019-02-11 17:24:28 INFO  AbstractConnector:318 - Stopped Spark@fff25f1{HTTP/1.1,[http/1.1]}{0.0.0.0:4040}
2019-02-11 17:24:28 INFO  SparkUI:54 - Stopped Spark web UI at http://master:4040
```

(6) 实验结果

本次实验结果均已给出。

3.4 Spark Python

拓展阅读

Python 介绍

1. 实验目的

了解 PySpark 的执行方式，学会使用 PyCharm 执行 PySpark。

2. 实验要求

成功配置 PyCharm 执行 PySpark 程序，完成 CountOnce 代码的编写。

CountOnce：已知一个数组，数组中只有一个数据是出现一次的，其他数据是出现两次的，将出现一次的数据找出。CountOnce 常用于数据块损坏检索，例如，HDFS 中每个数据块有两个副本，其中某一数据块出现损坏时，需要利用 CountOnce 找到损坏的数据块。[19]

例如，输入内容：

```
1 2 3 1 4 3 4 5 2 6 7 5 8 7 8
```

输出内容：

```
6
```

3. 预备知识

Python 是一种跨平台的计算机程序设计语言，是一种面向对象的动态类型语言，最初被设计用于编写自动化脚本(shell)，随着版本的不断更新和语言新功能的添加，越来越多地被用于独立的大型项目的开发。由于 Python 语言的简洁性、易读性以及可扩展性，在国外用 Python 做科学计算的研究机构日益增多，一些知名大学已经采用 Python 来教授程序设计课程，如卡耐基梅隆大学的编程基础、麻省理工学院的计算机科学及编程导论。众多开源的科学计算软件包都提供了 Python 的调用接口，如著名的计算机视觉库 OpenCV、三维可视化库 VTK、医学图像处理库 ITK。而 Python 专用的科学计算扩展库就更多了，如以下 3 个十分经典的科学计算扩展库：NumPy、SciPy 和 Matplotlib，它们分别为 Python 提供了快速数组处理、数值运算以及绘图功能。Python 语言及其众多的扩展库所构成的开发环境十分适合工程技术人员、科研人员处理实验数据、制作图表，甚至开发科学计算应用程序。

4. 实验环境

客户端环境：安装 PyCharm 和 Python 3.6.5。

服务端环境：需要 4 台安装 Linux 系统的物理机或虚拟机，本书使用 4 台 VirtualBox 虚拟机，配置如表 3-1 所示，需要注意的是，Spark 集群的 Python 版本需要和本地的版本一致，即 3.6.5。

5. 实验步骤

(1) PyCharm 配置

① 新建 Python 项目,建立 Python 文件 CountOnce. py。

② 打开 PyCharm 中的项目设置,选择"Project Structure",单击"＋Add Content Root",如图 3-22 所示。

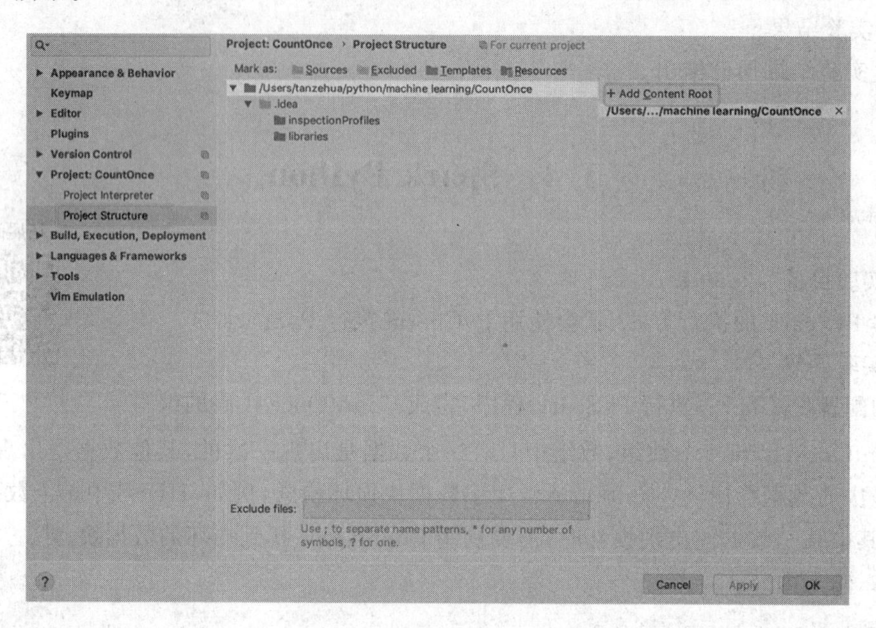

图 3-22　项目设置界面(一)

③ 添加. /spark-2. 3. 2-bin-hadoop2. 7/python/lib 目录下的"py4j-0. 10. 7-src. zip"文件和"pyspark. zip"文件(spark-2. 3. 2-bin-hadoop2. 7 需要自行解压到本地,不在集群上),单击"OK",如图 3-23 所示。

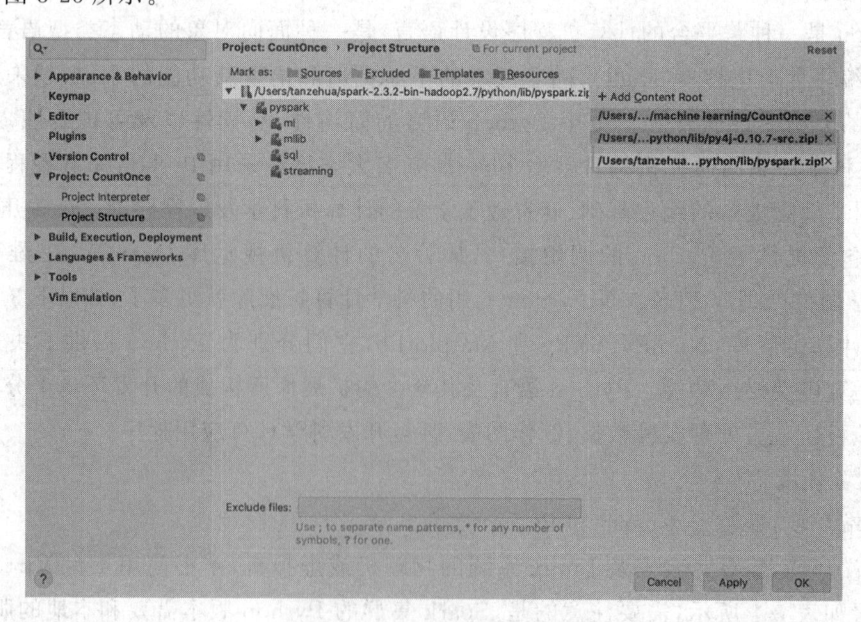

图 3-23　项目设置界面(二)

④ 右击“CountOnce. py”，选择“Create ‘CountOnce’”，如图 3-24 所示。

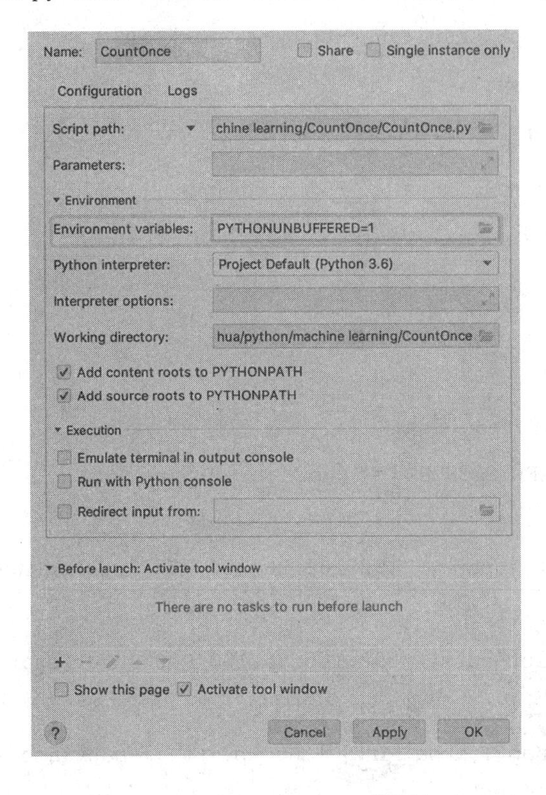

图 3-24　Create CountOnce 界面(一)

⑤ 选择“Environment variables”，添加 SPARK_HOME　/Users/tanzehua/spark-2. 3. 2-bin-hadoop2. 7(本地解压的 Spark 路径)，PYTHONPATH　/Users/tanzehua/spark-2. 3. 2-bin-hadoop2. 7/python(本地解压的 Spark 路径下的 Python 目录)，单击“OK”，如图 3-25 所示。

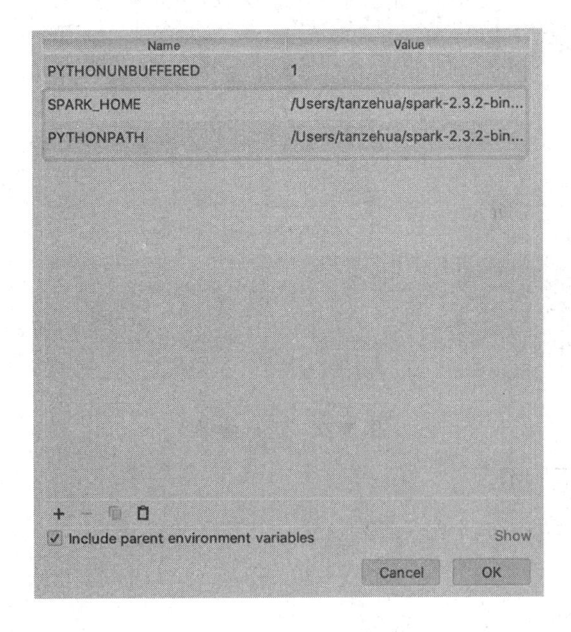

图 3-25　Create CountOnce 界面(二)

⑥ 上传输入文件至 HDFS：

```
hadoop fs -put -f /opt/test3_in.txt /user/spark/in/
```

文件内容：

```
1 2 3 1 4 3 4 5 2 6 7 5 8 7 8
```

(2) 编写 CountOnce 程序

① 代码编写如下所示。

```
import os
from pyspark import SparkContext


#指定集群使用的 Python 版本
os.environ["PYSPARK_PYTHON"] = "/usr/bin/python3"
#连接 Spark 集群
sc = SparkContext("spark://192.168.56.100:7077", "CountOnce")
#读取 HDFS 上的输入文件
arry = sc.textFile("hdfs://192.168.56.100:9000/user/spark/in/test3_in.txt")
#将文件内容划分为数字,并将 str 转为 int 类型
nums = arry.flatMap(lambda line: line.split('')).map(lambda x: int(x))
#对列表中编号做异或操作
result = nums.reduce(lambda x1, x2: x1 ^ x2)
#打印结果
print("*****num appear onece is :" + str(result) + "*****")
#关闭 SparkContext
sc.stop()
```

② 执行程序。

(3) 实验结果

① 执行结果如图 3-26 所示。

```
Setting default log level to "WARN".
To adjust logging level use sc.setLogLevel(newLevel). For SparkR, use setLogLevel(newLevel).
***** num appear onece is :6 *****

Process finis ed wit  exit code 0
```

图 3-26　运行结果

② 结果的内容如下所示。

```
***** num appear once is :6 *****
```

3.5 Spark MLlib

拓展阅读

机器学习介绍

1. 实验目的

了解 Spark MLlib 的基本使用方法。

2. 实验要求

利用 Spark MLlib，使用 Scala 语言编写线性回归预测程序。[20]

3. 预备知识

MLlib 是 Spark 中提供机器学习函数的库，它是专为在集群上并行运行的情况而设计的。MLlib 中包含许多机器学习算法，可以在 Spark 支持的所有编程语言中使用，由于 Spark 基于内存计算模型的优势，非常适合机器学习中出现的多次迭代，避免了操作磁盘和网络的性能损耗（Spark 官网展示的 MLlib 与 Hadoop 性能对比图就非常显著），因此 Spark 比 Hadoop 的 MapReduce 框架更易于支持机器学习。MLlib 支持的基础机器学习算法如表 3-2 所示。

表 3-2　MLlib 支持的基础机器学习算法

	离散型	连续型
有监督的机器学习	分类 逻辑回归 支持向量机(SVM) 朴素贝叶斯 决策树 随机森林 梯度提升决策树（GBT）	回归 线性回归 决策树 随机森林 梯度提升决策树（GBT） 保序回归
无监督的机器学习	聚类 k-means 高斯混合 快速迭代聚类(PIC) 隐含狄利克雷分布(LDA) 二分 k-means 流 k-means	协同过滤、降维 交替最小二乘(ALS) 奇异值分解(SVD) 主成分分析(PCA)

4. 实验环境

客户端环境：安装 IntelliJ IDEA，配置 Scala 开发环境，安装 Scala SDK 2.11 和 JDK 1.8。

服务端环境：需要 4 台安装 Linux 系统的物理机或虚拟机，本书使用 4 台 VirtualBox 虚拟机，配置如表 3-1 所示。

5. 实验步骤

(1) 建立项目

① 按照 3.3 节的内容建立 Maven＋Spark＋Scala 项目[21]，项目目录结构如图 3-27 所示。

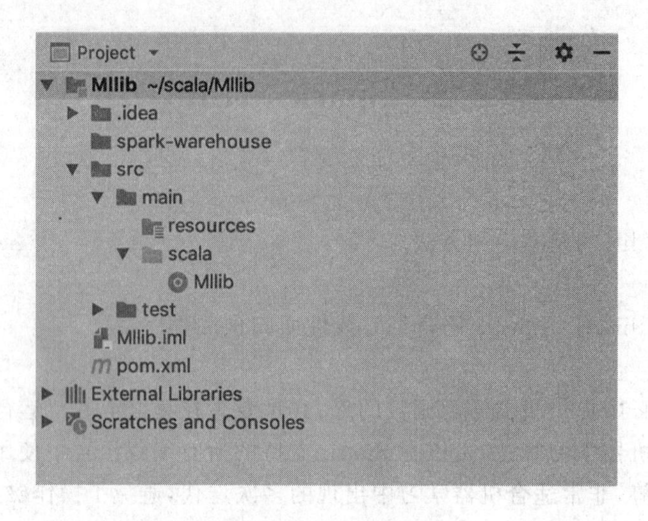

图 3-27 项目目录结构

② 修改 pom.xml 文件,内容如下所示。[22]

```
<? xml version = "1.0" encoding = "UTF-8"? >
<project xmlns = "http://maven.apache.org/POM/4.0.0"
         xmlns:xsi = "http://www.w3.org/2001/XMLSchema-instance"
         xsi:schemaLocation = "http://maven.apache.org/POM/4.0.0 http://maven.apache.
                               org/xsd/maven-4.0.0.xsd">
    <modelVersion>4.0.0</modelVersion>
    <properties>
        <spark.version>2.3.2</spark.version>
        <scala.version>2.11</scala.version>
    </properties>
    <groupId>com.bupt.pcncad</groupId>
    <artifactId>Mllib</artifactId>
    <version>1.0-SNAPSHOT</version>
    <dependencies>
    <dependency>
    <groupId>org.apache.spark</groupId>
    <artifactId>spark-core_${scala.version}</artifactId>
    <version>${spark.version}</version>
    </dependency>
        <dependency>
            <groupId>org.apache.spark</groupId>
            <artifactId>spark-mllib_${scala.version}</artifactId>
            <version>${spark.version}</version>
        </dependency>
```

```
    </dependencies>
    <build>
        <plugins>
          <plugin>
            <groupId>org.scala-tools</groupId>
            <artifactId>maven-scala-plugin</artifactId>
            <version>2.15.2</version>
            <executions>
              <execution>
                <goals>
                  <goal>compile</goal>
                  <goal>testCompile</goal>
                </goals>
              </execution>
            </executions>
          </plugin>
        </plugins>
    </build>
</project>
```

(2) 编写线性回归预测

① 导入数据(test4_in_tr.txt 文件)：

```
hadoop fs -put -p -f /opt/test4_in_tr.txt /user/spark/in
```

训练数据如下所示：

```
3615, 3624, 2.1, 69.05, 15.1, 41.3, 20, 50708
365, 6315, 1.5, 69.31, 11.3, 66.7, 152, 566432
2212, 4530, 1.8, 70.55, 7.8, 58.1, 15, 113417
2110, 3378, 1.9, 70.66, 10.1, 39.9, 65, 51945
21198, 5114, 1.1, 71.71, 10.3, 62.6, 20, 156361
2541, 4884, 0.7, 72.06, 6.8, 63.9, 166, 103766
3100, 5348, 1.1, 72.48, 3.1, 56, 139, 4862
579, 4809, 0.9, 70.06, 6.2, 54.6, 103, 1982
8277, 4815, 1.3, 70.66, 10.7, 52.6, 11, 54090
4931, 4091, 2, 68.54, 13.9, 40.6, 60, 58073
```

导入需要预测的测试数据(test4_in_pr.txt 文件)：

```
hadoop fs -put -p -f /opt/test4_in_pr.txt /user/spark/in
```

其中第 5 列为需要预测的测试数据：

```
4981, 4701, 1.4, 70.08, 9.5, 47.8, 85, 39780
3559, 4864, 0.6, 71.72, 4.3, 63.5, 32, 66570
1799, 3617, 1.4, 69.48, 6.7, 41.6, 100, 24070
4589, 4468, 0.7, 72.48, 3, 54.5, 149, 54464
376, 4566, 0.6, 70.29, 6.9, 62.9, 173, 97203
```

② 编写代码。

```
import org.apache.spark.ml.feature.VectorAssembler
import org.apache.spark.ml.regression.LinearRegression
import org.apache.spark.sql.{DataFrame, SQLContext}
import org.apache.spark.{SparkConf, SparkContext}

object Mllib {

def main(args: Array[String]){
  System.setProperty("HADOOP_USER_NAME", "root")
  val conf = new SparkConf().setAppName("Mllib").setMaster("spark://192.168.
            56.100:7077").setJars(List("./out/artifacts/Mllib_jar/Mllib.
            jar"))
  val sc = new SparkContext(conf)
  sc.setLogLevel("WARN")
  val sqc = new SQLContext(sc)
  // 读取训练数据
  val inputTr = sc.textFile("hdfs://192.168.56.100:9000/user/spark/in/test4_
                           in_tr.txt")
  // 将文件转化为元组形式
  val mapInputTr = inputTr.map{x =>
    val list = x.split(",")
    (list(0).toDouble, list(1).toDouble, list(2).toDouble,list(3).toDouble,
     list(4).toDouble, list(5).toDouble,list(6).toDouble,list(7).toDouble)
  }
  // 将 RDD 转为 DataFrame
  val inputDataFrameTr = sqc.createDataFrame(mapInputTr)
  // 添加 DataFrame 的列名
  val dataTr = inputDataFrameTr.toDF("Population", "Income", "Illiteracy",
                                     "LifeExp", "Murder", "HSGrad", "Frost",
                                     "Area")
  val colArray = Array("Population", "Income", "Illiteracy", "LifeExp",
                       "HSGrad", "Frost", "Area")
  // 设置列
```

```
val assemblerTr = new VectorAssembler().setInputCols(colArray).setOutputCol
            ("features")
// 向量列转化为 DataFrame
val vecDataFrameTr: DataFrame = assemblerTr.transform(dataTr)

// 读取预测数据
val inputPr = sc.textFile("hdfs://192.168.56.100:9000/user/spark/in/test4_
          in_pr.txt")
val mapInputPr = inputPr.map{x =>
  val split_list = x.split(",")
  (split_list(0).toDouble,split_list(1).toDouble,split_list(2).toDouble,
   split_list(3).toDouble,split_list(4).toDouble,split_list(5).toDouble,
   split_list(6).toDouble,split_list(7).toDouble)
  }
  // 将 RDD 转为 DataFrame
  val inputDataFramePr = sqc.createDataFrame(mapInputPr)
  // 添加 DataFrame 的列名
  val dataPr = inputDataFramePr.toDF("Population", "Income", "Illiteracy",
                          "LifeExp", " Murder ", " HSGrad ",
                          "Frost", "Area")
  val colArrayPr = Array("Population", "Income", "Illiteracy", "LifeExp",
                  "HSGrad", "Frost", "Area")
  // 设置列
   val assemblerPr = new VectorAssembler().setInputCols(colArrayPr).
                setOutputCol("features")
  // 向量列转化为 DataFrame
  val vecDataFramePr: DataFrame = assemblerPr.transform(dataPr)
  // 建立模型,预测 Murder
  val linearRegression1 = new LinearRegression()
  // setFeaturesCol 用于设置特征向量列名,setLabelCol 用于设置标签列名,
    setFitIntercept 用于设置是否计算截距
   val linearRegression2 = linearRegression1.setFeaturesCol("features").
                  setLabelCol("Murder").setFitIntercept(true)
  // setMaxIter 用于设置迭代次数,setRegParam 用于设置正则化参数
  val linearRegression3 = linearRegression2.setMaxIter(100).setRegParam(0.3).
                 setElasticNetParam(1)
  val linearRegression = linearRegression3
  // 将训练集合代入模型进行训练
  val lrModel = linearRegression.fit(vecDataFrameTr)
  // 输出模型结果
  println(s"系数: ${lrModel.coefficients} 截距: ${lrModel.intercept}")
  // 对模型进行评价
```

```
    val trainingSummary = lrModel.summary
    println(s"历史迭代损失：${trainingSummary.objectiveHistory.toList}")
    println(s"迭代次数：${trainingSummary.totalIterations}")
    println(s"均方根误差：${trainingSummary.rootMeanSquaredError}")
    println(s"决定系数 R2：${trainingSummary.r2}")
    val predictions：DataFrame = lrModel.transform(vecDataFramePr)
    println("输出预测结果")
    predictions.selectExpr("features","Murder", "round(prediction,1)
                          as prediction").show(false)
    sc.stop()
  }
}
```

（3）实验结果

① 执行结果如图 3-28 所示。

```
系数: [1.8823762050845994E-4,-3.6225420647245844E-4,1.5390509779556736,-1.6599231598034005,0.0,0.0,1.364196112858076E-6] 截距:
 124.97141181368359
历史迭代损失:List(0.5000000000000013, 0.4537585109029017, 0.22623894231713423, 0.21513615533886027, 0.20690491009903425, 0
 .20153409895059557, 0.19881427873854202, 0.19662933682038788, 0.19658562994287232, 0.19619406201887296, 0.1961126236317578,
 0.19601282275880078, 0.19596502879006752, 0.19593690605019673, 0.19592341951975717, 0.19591496427259994, 0.19590601645628744,
 0.19589500816236688, 0.19587448463882984, 0.1958665332472571, 0.19585711911276174, 0.19585208071016033, 0.19584609882204987,
 0.19584455572914433, 0.19584429885724403, 0.19584393904790434, 0.19584388796842442, 0.19584388010959908, 0.19584387892530097,
 0.19584387871873413, 0.19584387868462041, 0.19584387867604816, 0.19584387867088157, 0.19584387866987424, 0.1958438786695783,
 0.19584387866913539, 0.19584387866929434, 0.1958438786691881, 0.19584387866921046)
迭代次数: 39
均方根误差: 1.416093145176352
决定系数R2: 0.8306618086474988
输出预测结果
+---------------------------------------+------+----------+
|features                               |Murder|prediction|
+---------------------------------------+------+----------+
|[4981.0,4701.0,1.4,70.08,47.8,85.0,39780.0] |9.5   |10.1      |
|[3559.0,4864.0,0.6,71.72,63.5,32.0,66570.0] |4.3   |5.8       |
|[1799.0,3617.0,1.4,69.48,41.6,100.0,24070.0]|6.7   |10.9      |
|[4589.0,4468.0,0.7,72.48,54.5,149.0,54464.0]|3.0   |5.1       |
|[376.0,4566.0,0.6,70.29,62.9,173.0,97203.0] |6.9   |7.8       |
+---------------------------------------+------+----------+

Process finished with exit code 0
```

图 3-28　执行结果

② 结果内容如下所示。

系数：[1.8823762050845994E-4,-3.6225420647245844E-4,1.5390509779556736,-1.6599231598034005,0.0,0.0,1.364196112858076E-6] 截距：124.97141181368359

历史迭代损失:List(0.5000000000000013，0.4537585109029017，0.22623894231713423，0.21513615533886027，0.20690491009903425，0.20153409895059557，0.19881427873854202，0.19662933682038788，0.19658562994287232，0.19619406201887296，0.1961126236317578，0.19601282275880078，0.19596502879006752，0.19593690605019673，0.19592341951975717，0.19591496427259994，0.19590601645628744，0.19589500816236688，0.19587448463882984，0.1958665332472571，0.19585711911276174，0.19585208071016033，0.19584609882204987，0.19584455572914433，0.19584429885724403，0.19584393904790434，0.19584388796842442，0.19584388010959908，0.19584387892530097，0.19584387871873413，0.19584387868462041，0.19584387867604816，0.19584387867088157，0.19584387866987424，0.1958438786695783，

```
0.19584387866913539，0.19584387866929434，0.1958438786691881，0.19584387866921046)
迭代次数：39
均方根误差：1.416093145176352
决定系数R2：0.8306618086474988
输出预测结果
```

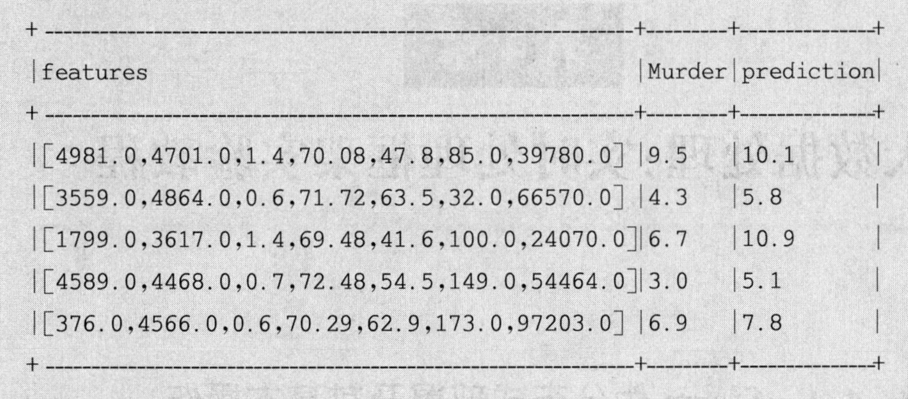

```
+--------------------------------------------------------+--------+------------+
|features                                                |Murder|prediction|
+--------------------------------------------------------+--------+------------+
|[4981.0,4701.0,1.4,70.08,47.8,85.0,39780.0] |9.5    |10.1       |
|[3559.0,4864.0,0.6,71.72,63.5,32.0,66570.0] |4.3    |5.8        |
|[1799.0,3617.0,1.4,69.48,41.6,100.0,24070.0]|6.7    |10.9       |
|[4589.0,4468.0,0.7,72.48,54.5,149.0,54464.0]|3.0    |5.1        |
|[376.0,4566.0,0.6,70.29,62.9,173.0,97203.0] |6.9    |7.8        |
+--------------------------------------------------------+--------+------------+

Process finished with exit code 0
```

第4章

大数据处理:实时处理框架实验教程

4.1　Storm 伪分布式部署及其基本操作

1. 实验目的

学会安装实时处理框架 Storm,掌握其基本操作。

2. 实验要求

① Java 8。

② Linux、MacOSX 或其他类似于 Unix 的操作系统(不支持 Windows)。

③ 8 GB 内存。

④ 2 个 vCPU。

⑤ ZooKeeper 分布式协调服务。

⑥ Maven。

⑦ Eclipse 或 IDEA 等编辑器。

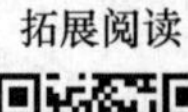

3. 预备知识

Storm 介绍

Apache Storm 是自由开源的分布式实时计算系统,擅长处理海量数据,适用于数据实时处理而非批处理。Storm 运行在 Java 虚拟机上且其主要部分是由 Java 和 Clojure 编写的,而守护进程及管理命令是由 Python 编写的,所以运行 Storm 需要具备 Java 和 Python 两种环境。Storm 还是一个支持多语言开发的框架,其中的 Spout 和 Bolt 部分可以使用多种支持数据流读取的编程语言编写。

Storm 有本地模式和集群模式两种运行模式。在本地模式下,主要是对 Storm 进行开发及测试。在集群模式下,主要是将开发好的程序 jar 包部署到 Storm 集群中运行。

4. 实验内容[23]

(1) 实验 1:Storm 伪分布式安装

① 下载 apache-storm-1.2.2.tar.gz 安装包。

② 终端运行 tar 命令解压安装包:

```
tar -zxvf /install/apache-storm-1.2.2.tar.gz
```

③ 启动 ZooKeeper：

```
zkServer.sh start
```

④ 启动 Storm 并确认安装成功：

```
storm nimbus &
storm ui &
storm supervisor &
```

启动成功后查看 Java 进程：

```
jps
```

以下是安装成功后的结果：

```
[root@hservermain ~]# jps
2977 nimbus
3046 core
3593 Jps
3005 Supervisor
2958 QuorumPeerMain
```

在浏览器页面查看 Storm 运行的详细信息：在浏览器中输入 IP:Port，如 192.168.1.2:8080，其中 8080 为 Storm 默认端口。

(2) 实验 2：Topology 作业的两种运行方式

① Storm 官方提供了入门代码 WordCount，但是其中包含 Python 文件，为了保证运行的简便，可使用修改后的示例代码(见扩展资料)。

② 使用 Eclipse 或 IDEA 等项目开发软件创建一个新的 Maven 项目，将 Storm 的核心依赖添加到 pom.xml 中。

```
<dependency>
    <groupId>org.apache.storm</groupId>
    <artifactId>storm-core</artifactId>
    <version>1.2.2</version>
    <scope>provided</scope>
</dependency>
```

③ 本地运行：将示例代码复制到 Maven 项目中，并运行。

④ 集群运行：将运行成功的代码打包，并发布到服务器端运行，通过浏览器方式查看作业运行情况。

5. 实验作业

① 下载 Storm 安装包并在虚拟机上安装。

② 在本地模式和集群模式两种模式下运行 Storm 示例 WordCount。

6. 扩展资料

(1) 实验 1：Storm 伪分布式安装

① 下载 apache-storm-1.2.2.tar.gz 安装包，并解压，如图 4-1 所示。

```
[root@hservermain software]# tar -zxvf /install/apache-storm-1.2.2.tar.gz
```

图 4-1 解压安装文件

解压后会出现蓝色的文件，如图 4-2 所示。

```
[root@hservermain software]# ls
apache-storm-1.2.2
[root@hservermain software]#
```

图 4-2 查看安装结果

② 因为是伪分布式环境，所以无须配置。在实际生产时，搭建分布式环境需要修改 Storm 主目录下 conf/storm. yaml 文件的相关信息，具体配置方法参考官方文档，地址为 http://storm. apache. org/releases/current/Configuration. html。

③ 启动 ZooKeeper 与 Storm 相关进程。

启动 ZooKeeper，如图 4-3 所示。

```
zkServer.sh start
```

```
[root@hservermain bin]# ./zkServer.sh start
ZooKeeper JMX enabled by default
Using config: /software/zookeeper-3.4.9/bin/../conf/zoo.cfg
Starting zookeeper ... STARTED
[root@hservermain bin]#
```

图 4-3 开启 ZooKeeper 服务

启动 Storm 相关进程，如图 4-4 所示。由于需要读取配置文件等，启动后需要等待一分钟左右。

```
storm nimbus &
storm supervisor &
storm ui &
```

```
[root@hservermain software]# storm nimbus &
[1] 9590
[root@hservermain software]# storm supervisor &
[2] 9616
[root@hservermain software]# storm ui &
[3] 9653
[root@hservermain software]#
```

图 4-4 启动 Storm 相关进程

全部启动成功后使用 jps 命令查看，如图 4-5 所示。

```
[root@hservermain software]# jps
9616 Supervisor
9537 QuorumPeerMain
9842 Jps
9653 core
9590 nimbus
```

图 4-5 进程列表

④ 在浏览器页面输入“machine IP:8080”即可查看 Storm 相关节点的工作状态。例如，输入 192.168.159.130:8080，查看到的状态如图 4-6 所示。

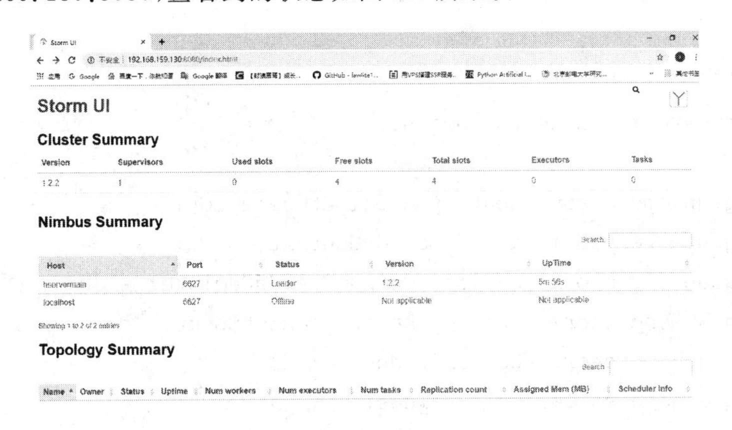

图 4-6 集群管理页

（2）实验 2：Topology 作业的两种运行方式

① 在 Eclipse 中创建一个新的 Maven 项目，项目目录如图 4-7 所示。

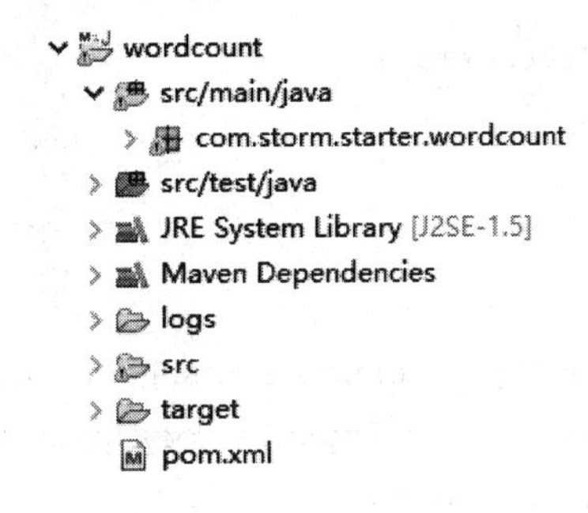

图 4-7 创建项目

② 在 pom.xml 中加入 Storm 项目开发的核心依赖，修改后 pom.xml 的内容如图 4-8 所示。如果下面本地运行时出现找不到 Storm 核心包中的接口或类的情况，将这里 storm-core 依赖包的<scope>项注释掉即可。

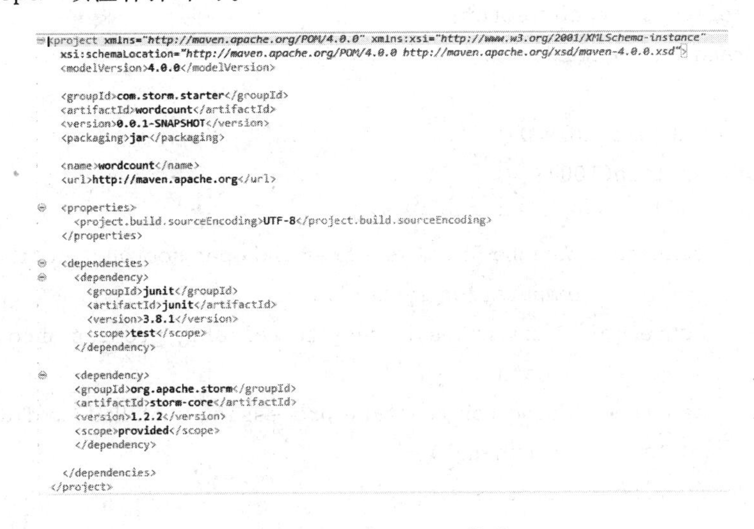

```xml
<project xmlns="http://maven.apache.org/POM/4.0.0" xmlns:xsi="http://www.w3.org/2001/XMLSchema-instance"
  xsi:schemaLocation="http://maven.apache.org/POM/4.0.0 http://maven.apache.org/xsd/maven-4.0.0.xsd">
  <modelVersion>4.0.0</modelVersion>

  <groupId>com.storm.starter</groupId>
  <artifactId>wordcount</artifactId>
  <version>0.0.1-SNAPSHOT</version>
  <packaging>jar</packaging>

  <name>wordcount</name>
  <url>http://maven.apache.org</url>

  <properties>
    <project.build.sourceEncoding>UTF-8</project.build.sourceEncoding>
  </properties>

  <dependencies>
    <dependency>
      <groupId>junit</groupId>
      <artifactId>junit</artifactId>
      <version>3.8.1</version>
      <scope>test</scope>
    </dependency>

    <dependency>
    <groupId>org.apache.storm</groupId>
    <artifactId>storm-core</artifactId>
    <version>1.2.2</version>
    <scope>provided</scope>
    </dependency>

  </dependencies>
</project>
```

图 4-8 添加 Maven 依赖

③ 在 com. storm. starter. wordcount 下创建两个 Java 文件，名为 RandomSentenceSpout 和 WordCountTopology，将以下代码粘贴进去。对于该案例的代码解析请参考相关资料，这里不再赘述。

RandomSentenceSpout. java：

```
package com.storm.starter.wordcount;

import org.apache.storm.spout.SpoutOutputCollector;
import org.apache.storm.task.TopologyContext;
import org.apache.storm.topology.OutputFieldsDeclarer;
import org.apache.storm.topology.base.BaseRichSpout;
import org.apache.storm.tuple.Fields;
import org.apache.storm.tuple.Values;
import org.apache.storm.utils.Utils;
import org.slf4j.Logger;
import org.slf4j.LoggerFactory;

import java.text.SimpleDateFormat;
import java.util.Date;
import java.util.Map;
import java.util.Random;

public class RandomSentenceSpout extends BaseRichSpout {
    private static final Logger LOG = LoggerFactory.getLogger(RandomSentenceSpout.class);
    SpoutOutputCollector _collector;
    Random _rand;

    public void open(Map conf, TopologyContext context, SpoutOutputCollector
                    collector) {
        _collector = collector;
        _rand = new Random();
    }
    public void nextTuple() {
        Utils.sleep(100);
        String[] sentences = new String[]{
            sentence("Apache Storm is a free and open source distributed realtime
                    computation system"),
            sentence("Storm makes it easy to reliably process unbounded streams
                    of data"),
            sentence("doing for realtime processing what Hadoop did for batch
                    processing"),
            sentence("Storm is simple"),
```

```
            sentence("can be used with any programming language")};
        final String sentence = sentences[_rand.nextInt(sentences.length)];
        LOG.debug("Emitting tuple: {}", sentence);
        _collector.emit(new Values(sentence));
    }
    protected String sentence(String input) {
      return input;
    }

    @Override
    public void ack(Object id) {
    }

    @Override
    public void fail(Object id) {
    }

    public void declareOutputFields(OutputFieldsDeclarer declarer) {
      declarer.declare(new Fields("word"));
    }

    public static class TimeStamped extends RandomSentenceSpout {
      private final String prefix;

      public TimeStamped() {
        this("");
      }

      public TimeStamped(String prefix) {
        this.prefix = prefix;
      }

      protected String sentence(String input) {
        return prefix + currentDate() + " " + input;
      }

      private String currentDate() {
         return new SimpleDateFormat("yyyy.MM.dd_HH:mm:ss.SSSSSSSSS").format
                                         (new Date());
      }
    }
}
```

WordCountTopology.java：

```
package com.storm.starter.wordcount;

import org.apache.storm.Config;
import org.apache.storm.LocalCluster;
import org.apache.storm.StormSubmitter;
import org.apache.storm.task.OutputCollector;
import org.apache.storm.task.TopologyContext;
import org.apache.storm.topology.BasicOutputCollector;
import org.apache.storm.topology.IRichBolt;
import org.apache.storm.topology.OutputFieldsDeclarer;
import org.apache.storm.topology.TopologyBuilder;
import org.apache.storm.topology.base.BaseBasicBolt;
import org.apache.storm.tuple.Fields;
import org.apache.storm.tuple.Tuple;
import org.apache.storm.tuple.Values;

import java.util.ArrayList;
import java.util.Collections;
import java.util.HashMap;
import java.util.List;
import java.util.Map;

public class WordCountTopology {
  public static class SplitSentence implements IRichBolt {
    private OutputCollector _collector;
    public void declareOutputFields(OutputFieldsDeclarer declarer) {
      declarer.declare(new Fields("word"));
    }

    public Map<String, Object> getComponentConfiguration() {
      return null;
    }

    public void prepare(Map stormConf, TopologyContext context,OutputCollector
                        collector) {
        _collector = collector;
    }

    public void execute(Tuple input) {
```

```
            String sentence = input.getStringByField("word");
            String[] words = sentence.split(" ");
            for(Stringword : words){
                this._collector.emit(new Values(word));
            }
        }

        public void cleanup() {
            // TODO Auto-generated method stub
        }
    }

    public static class WordCount extends BaseBasicBolt {
        Map<String, Integer> counts = new HashMap<String, Integer>();

        public void execute(Tuple tuple, BasicOutputCollector collector) {
            String word = tuple.getString(0);
            Integer count = counts.get(word);
            if (count == null)
                count = 0;
            count++;
            counts.put(word, count);
            collector.emit(new Values(word, count));
        }

        public void declareOutputFields(OutputFieldsDeclarer declarer) {
            declarer.declare(new Fields("word", "count"));
        }
    }

    public static class WordReport extends BaseBasicBolt {
            Map<String, Integer> counts = new HashMap<String, Integer>();

            public void execute(Tuple tuple, BasicOutputCollector collector) {
                String word = tuple.getStringByField("word");
                Integer count = tuple.getIntegerByField("count");
                this.counts.put(word, count);
            }
            public void declareOutputFields(OutputFieldsDeclarer declarer) {

            }

            @Override
            public void cleanup() {
```

```
            System.out.println("----------------FINAL COUNTSSTART----------------------");
            List<String> keys = new ArrayList<String>();
            keys.addAll(this.counts.keySet());
            Collections.sort(keys);

            for(String key : keys){
                System.out.println(key + " : " + this.counts.get(key));
            }

            System.out.println("----------------FINAL COUNTSEND----------------------");
        }
    }
    public static void main(String[] args) throws Exception {
      TopologyBuilder builder = new TopologyBuilder();
      builder.setSpout("spout", new RandomSentenceSpout(), 5);
      builder.setBolt("split", new SplitSentence(), 8).shuffleGrouping("spout");
      builder.setBolt("count", new WordCount(), 12).fieldsGrouping("split", new
                    Fields("word"));
      builder.setBolt("report", new WordReport(), 6).globalGrouping("count");
      Config conf = new Config();
      conf.setDebug(true);

      if (args != null && args.length > 0) {
        conf.setNumWorkers(3);

        StormSubmitter.submitTopologyWithProgressBar(args[0], conf, builder.
                                                createTopology());
      }
      else {
        conf.setMaxTaskParallelism(3);
        LocalCluster cluster = new LocalCluster();
        cluster.submitTopology("word-count", conf, builder.createTopology());

        Thread.sleep(20000);

        cluster.shutdown();
      }
    }
}
```

④ 主方法在 WordCountTopology 中，运行程序，如图 4-9 所示。

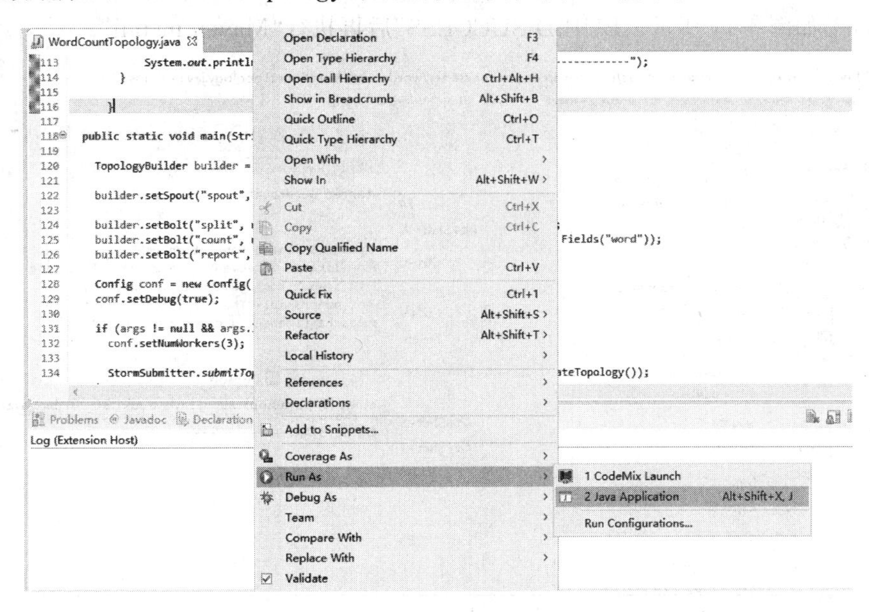

图 4-9　程序运行

运行后中途出现图 4-10 所示的输出结果，代表运行成功。如果未出现以下结果，则和代码设置的进程睡眠时间以及机器运行速度有关，多运行几次即可。

```
-----------------FINAL COUNTS  START-----------------------
Apache : 46
Hadoop : 38
Storm : 106
a : 46
and : 46
any : 27
batch : 38
be : 27
can : 27
computation : 46
data : 30
did : 38
distributed : 46
doing : 38
easy : 30
for : 76
free : 46
is : 76
it : 30
language : 27
makes : 30
of : 30
open : 46
process : 30
processing : 76
programming : 27
realtime : 84
reliably : 30
simple : 30
source : 46
streams : 30
system : 46
to : 30
unbounded : 30
used : 27
what : 38
```

图 4-10　程序运行结果

⑤ 将运行成功的项目打包,发布到 Storm 集群运行,具体操作为右击项目,选择“Run As→Maven clean”,如图 4-11 所示,“BUILD SUCCESS”后再执行“Maven install”。

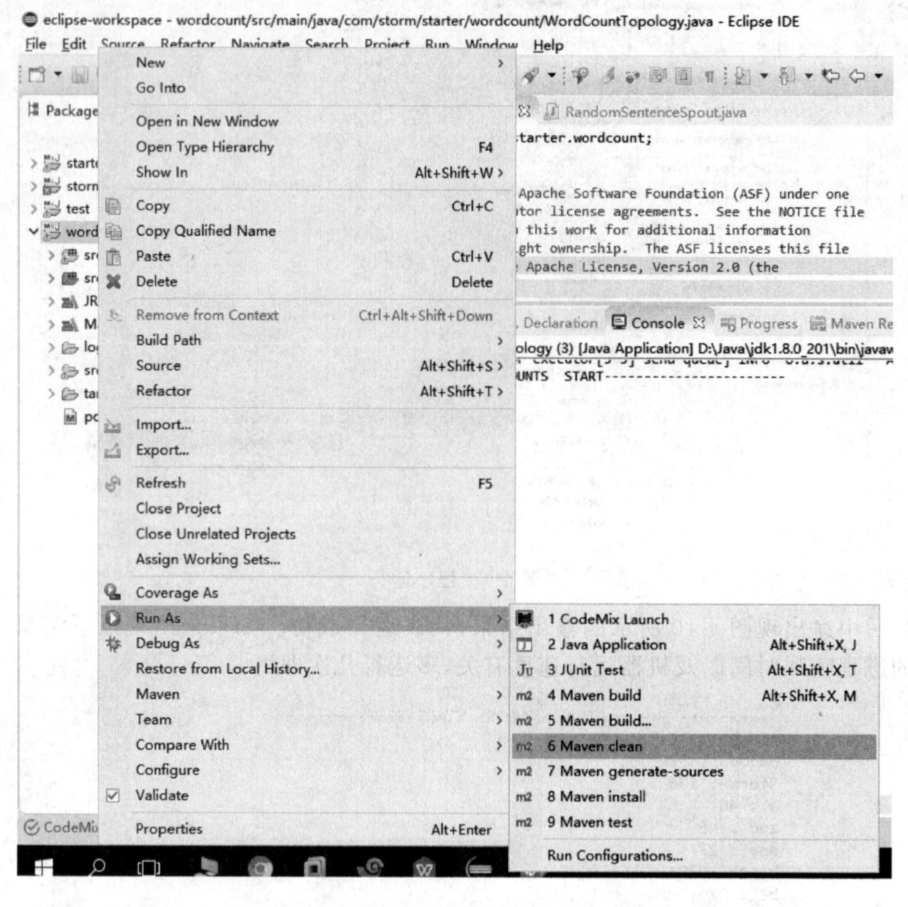

图 4-11 程序打包

“Maven install”成功后在项目目录 target 下会生成打包好的文件,名为 wordcount-0. 0. 1-SNAPSHOT. jar。将项目 jar 包通过 Xftp 传到服务器端某一目录下,如图 4-12 所示。

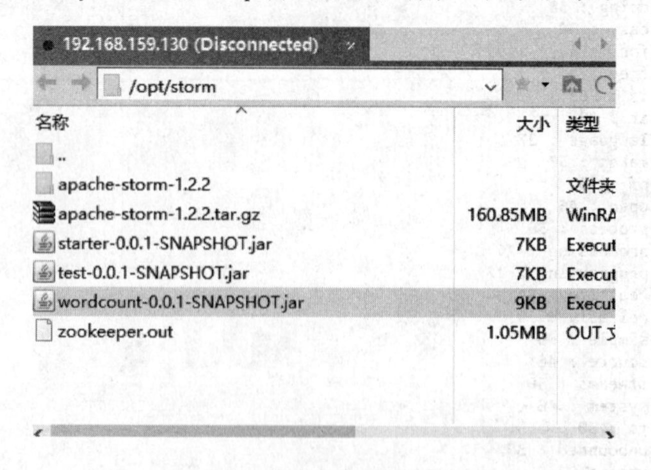

图 4-12 将程序上传到服务器

⑥ 执行以下命令,如图 4-13 所示。其中,jar 后面的是项目包所在路径、项目主方法所在

类以及发布到 Storm 运行的拓扑名称(自定义)。

```
storm jar ./wordcount-0.0.1-SNAPSHOT.jar
com.storm.starter.wordcount.WordCountTopology word-count
```

```
[root@hservermain software]# storm jar ./wordcount-0.0.1-SNAPSHOT.jar com.storm.starter.wordcount.WordCount
Topology word-count

Running: /software/java/bin/java -client -Ddaemon.name= -Dstorm.options= -Dstorm.home=/software/apache-stor
m-1.2.2 -Dstorm.log.dir=/software/apache-storm-1.2.2/logs -Djava.library.path=/usr/local/lib:/opt/local/lib
:/usr/lib -Dstorm.conf.file= -cp /software/apache-storm-1.2.2/*:/software/apache-storm-1.2.2/lib/*:/softwar
e/apache-storm-1.2.2/extlib/*:./wordcount-0.0.1-SNAPSHOT.jar:/software/apache-storm-1.2.2/conf:/software/ap
ache-storm-1.2.2/bin -Dstorm.jar=./wordcount-0.0.1-SNAPSHOT.jar -Dstorm.dependency.jars= -Dstorm.dependency
.artifacts={} com.storm.starter.wordcount.WordCountTopology word-count
```

图 4-13　提交作业到 Storm 集群

可以看到作业已经成功提交到 Storm 集群,如图 4-14 所示。

```
5350 [main] INFO  o.a.s.StormSubmitter - Generated ZooKeeper secret payload for MD5-digest: -75515888193596
62110:-9043542814401335017
6547 [main] INFO  o.a.s.u.NimbusClient - Found leader nimbus : hservermain:6627
6702 [main] INFO  o.a.s.s.a.AuthUtils - Got AutoCreds []
6722 [main] INFO  o.a.s.u.NimbusClient - Found leader nimbus : hservermain:6627
6961 [main] INFO  o.a.s.StormSubmitter - Uploading dependencies - jars...
6963 [main] INFO  o.a.s.StormSubmitter - Uploading dependencies - artifacts...
6963 [main] INFO  o.a.s.StormSubmitter - Dependency Blob keys - jars : [] / artifacts : []
7080 [main] INFO  o.a.s.StormSubmitter - Uploading topology jar ./wordcount-0.0.1-SNAPSHOT.jar to assigned
location: /software/apache-storm-1.2.2/storm-local/nimbus/inbox/stormjar-72bf89ee-ac9d-4eb5-beal-cdc81f5cc2
67.jar
Start uploading file './wordcount-0.0.1-SNAPSHOT.jar' to '/software/apache-storm-1.2.2/storm-local/nimbus/i
nbox/stormjar-72bf89ee-ac9d-4eb5-beal-cdc81f5cc267.jar' (9606 bytes)
[==================================================] 9606 / 9606
File './wordcount-0.0.1-SNAPSHOT.jar' uploaded to '/software/apache-storm-1.2.2/storm-local/nimbus/inbox/st
ormjar-72bf89ee-ac9d-4eb5-beal-cdc81f5cc267.jar' (9606 bytes)
7254 [main] INFO  o.a.s.StormSubmitter - Successfully uploaded topology jar to assigned location: /software
/apache-storm-1.2.2/storm-local/nimbus/inbox/stormjar-72bf89ee-ac9d-4eb5-beal-cdc81f5cc267.jar
7254 [main] INFO  o.a.s.StormSubmitter - Submitting topology word-count in distributed mode with conf {"sto
rm.zookeeper.topology.auth.scheme":"digest","storm.zookeeper.topology.auth.payload":"-7551588819359662110:-
9043542814401335017","topology.workers":3,"topology.debug":true}
7254 [main] WARN  o.a.s.u.Utils - STORM-VERSION new 1.2.2 old 1.2.2
8279 [main] INFO  o.a.s.StormSubmitter - Finished submitting topology: word-count
[root@hservermain software]#
```

图 4-14　提交作业到 Storm 集群成功

⑦ 在浏览器内查看集群的工作状态,可以发现此时 Topology 中已经多出一个正在运行的作业,名为 word-count,如图 4-15 所示。

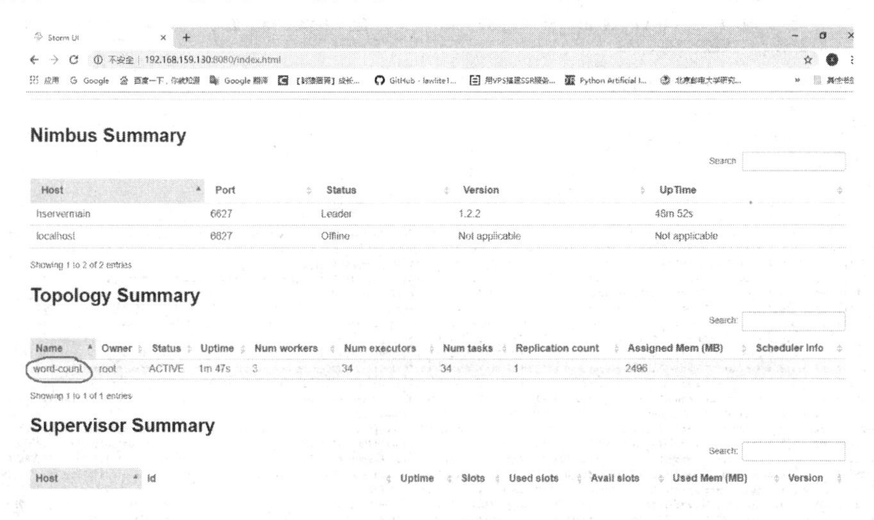

图 4-15　集群管理页

点击 word-count 作业,可以查看作业运行的相关信息,其中 worker 的数量、executor 的数量以及 task 的数量是在程序中设置的。

⑧ 启动 logviewer 进程才可以看到作业工作的日志信息，使用命令“storm logviewer &”启动日志进程，如图 4-16 所示。

```
[root@hservermain software]# storm logviewer &
[4] 2368
```

图 4-16 logviewer 启动

启动后使用 jps 命令可以查看到有 logviewer 进程，如图 4-17 所示。

```
[root@hservermain software]# jps
2368 logviewer
1315 core
1160 nimbus
2504 Jps
1258 QuorumPeerMain
1467 Supervisor
```

图 4-17 进程列表

在“Worker Resources”中可以查看作业日志信息，单击端口进行日志查看，如图 4-18 所示。

Worker Resources

Search: [ ] Toggle Components

Host	Supervisor Id	Port	Uptime	Num executors	Assigned Mem (MB)	Components
hservermain	b1710e42-c53b-4c84-9feb-4dcfc4356527	6701	2m 30s	11	832	4 components
hservermain	b1710e42-c53b-4c84-9feb-4dcfc4356527	6702	2m 36s	11	832	4 components
hservermain	b1710e42-c53b-4c84-9feb-4dcfc4356527	6700	2m 34s	12	832	4 components

图 4-18 Worker 日志信息

程序运行日志如图 4-19 所示。

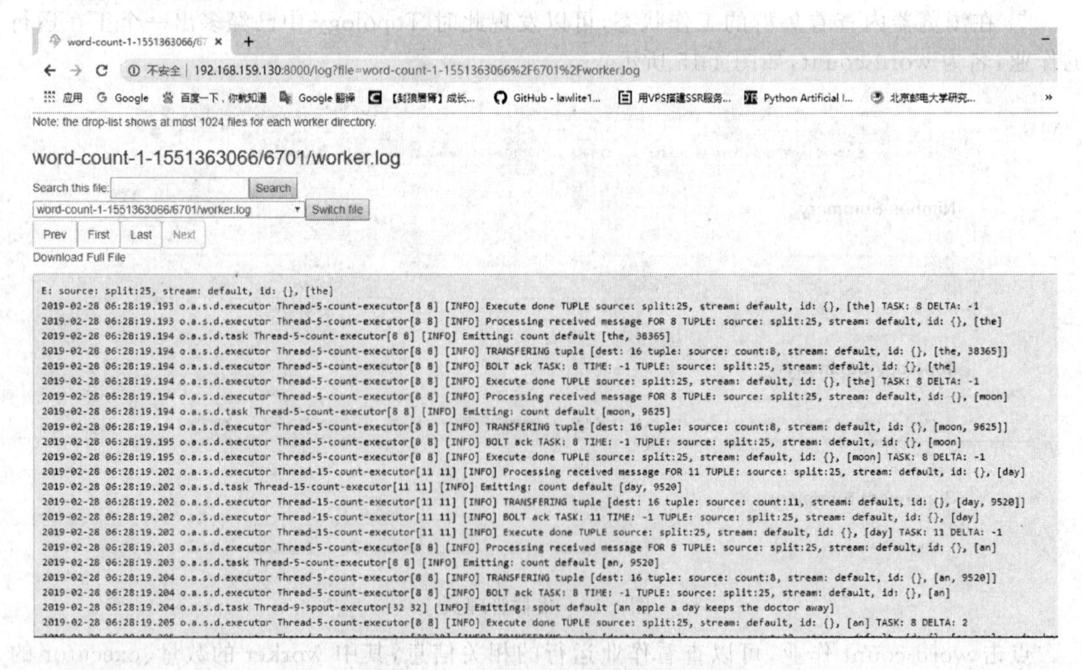

图 4-19 word-count 程序运行日志

⑨ 由于 Storm 的特性，作业一旦启动，除非手动停止，否则会一直运行。这一点可以通过观察 Topology 状态查看，时间正常，Emitted 和 Transferred 数量持续增加，说明程序一直在运行，如图 4-20 所示。

Topology stats

Window	Emitted	Transferred
10m 0s	409902	409870
3h 0m 0s	797040	797040
1d 0h 0m 0s	797040	797040
All time	797040	797040

图 4-20　Topology 状态

重启服务器后，重新开启 ZooKeeper 以及 Storm，在浏览器中会发现重启前提交的 word-count 作业依然处于运行状态，如图 4-21 所示。

Topology summary

Name	Id	Owner	Status	Uptime	Num workers	Num executors	Num tasks	Replication count	Assigned Mem (MB)	Scheduler info
word-count	word-count-1-1551363066	root	ACTIVE	28m 29s	3	34	34	1	2496	

图 4-21　Topology 列表

通过 Topology 管理页面的 kill 选项或者在服务器端命令行使用"storm kill word-count"可以停止作业的运行，如图 4-22 和 4-23 所示。

图 4-22　管理页停止 Topology

```
[root@hservermain software]# storm kill word-count

Running: /software/java/bin/java -client -Ddaemon.name= -Dstorm.options= -Dstorm.home=/software/apache-stor
m-1.2.2 -Dstorm.log.dir=/software/apache-storm-1.2.2/logs -Djava.library.path=/usr/local/lib:/opt/local/lib
:/usr/lib -Dstorm.conf.file= -cp /software/apache-storm-1.2.2/*:/software/apache-storm-1.2.2/lib/*:/softwar
e/apache-storm-1.2.2/extlib/*:/software/apache-storm-1.2.2/extlib-daemon/*:/software/apache-storm-1.2.2/con
f:/software/apache-storm-1.2.2/bin org.apache.storm.command.kill_topology word-count
```

图 4-23　使用命令停止 Topology

如图 4-24 所示，此时集群中没有作业运行。

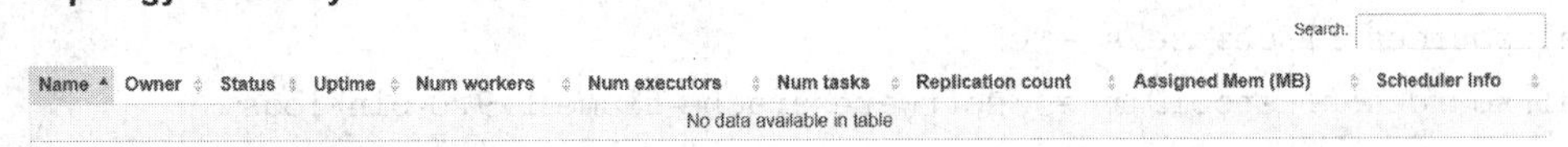

Topology Summary

Search:

Name	Owner	Status	Uptime	Num workers	Num executors	Num tasks	Replication count	Assigned Mem (MB)	Scheduler Info
No data available in table									

图 4-24　停止 Topology 后的列表

7. 参考答案

实验作业的答案见本节扩展资料部分。

4.2 Flume 安装及其基本操作

1. 实验目的

学会在计算机上安装实时数据收集系统 Flume，掌握其获取数据并导入 HDFS 的流程。

2. 实验要求

① Java 8。

② Linux、MacOSX 或其他类似于 Unix 的操作系统（不支持 Windows）。

③ 8 GB 内存。

④ 2 个 vCPU。

⑤ Hadoop。

3. 预备知识

Flume 是一个分布式、可靠性高、可用性高的海量数据采集、聚合和传输系统，它将大量数据从许多不同的源移动到存储系统（如 HDFS）中集中存储，类似于物理上的一根数据线。

Flume 最重要的三个组件是 Source、Channel、Sink，Source 用于获取数据，Channel 是通道，用于传输以及缓存数据，Sink 用于数据推送。一个 Flume 服务主要是针对这三个组件的配置文件启动的。

4. 实验内容[24]

（1）实验 1：Flume 安装

① 下载 apache-flume-1.9.0-bin.tar.gz 安装包。

② 终端运行 tar 命令解压安装包。

```
tar -zxvf /install/apache-flume-1.9.0-bin.tar.gz
```

③ 复制配置文件，重命名为 flume-env.sh，并修改相关配置。

④ 验证安装是否成功。

```
bin/flume-ng version
```

（2）实验 2：Flume 获取数据并导入 HDFS

① 创建 agent 配置文件，文件名为 spool.conf。

```
a1.sources = r1
a1.sinks = k1
a1.channels = c1

# Describe/configure the source
a1.sources.r1.type = spooldir
a1.sources.r1.channels = c1
a1.sources.r1.spoolDir = /software/apache-flume-1.9.0-bin/logs
a1.sources.r1.fileHeader = true

# Describe the sink
a1.sinks.k1.type = hdfs
a1.sinks.k1.hdfs.path = hdfs://hservermain:9000/flume/%y-%m-%d/%H%M
```

```
a1.sinks.k1.hdfs.useLocalTimeStamp = true
a1.sinks.k1.hdfs.round = true
a1.sinks.k1.hdfs.roundValue = 10
a1.sinks.k1.hdfs.roundUnit = minute

# Use a channel which buffers events in memory
a1.channels.c1.type = memory
a1.channels.c1.capacity = 1000
a1.channels.c1.transactionCapacity = 100

# Bind the source and sink to the channel
a1.sources.r1.channels = c1
a1.sinks.k1.channel = c1
```

② 创建数据收集源目录/software/apache-flume-1.9.0-bin/logs。

```
mkdir /software/apache-flume-1.9.0-bin/logs
```

③ 启动 Hadoop 集群。

```
start-all.sh
```

④ 启动 Flume。

```
bin/flume-ng agent -n a1 -c conf -f conf/spool.conf -Dflume.root.logger = INFO,
console
```

⑤ 在当前虚拟机中重新打开一个窗口，编辑一个测试文件 test.log，编辑内容如下所示。

```
Apache Flume is adistributed, reliable, and available
system for efficiently collecting, aggregating and moving
large amounts of log data from many different sources to a centralized data store.
```

⑥ 将测试文件复制到数据收集源目录。

```
cp test.log logs/
```

⑦ 查看 Flume 进程实时输出的日志信息，在 HDFS 上查看生成的文件并下载到本地。

（3）实验 3：完成其他 Flume 案例

实验 3 可参考官方文档中对众多不同的 Source、Channel、Sink 的详细说明及其他参考资料完成。官网：http://flume.apache.org/releases/content/1.9.0/FlumeUserGuide.html。

5. 实验作业

① 下载 Flume 安装包并在虚拟机上安装。

② 使用 Flume 实时获取数据并导入 HDFS。

③ 使用 Flume 完成其他应用案例（要求不能使用与实验 2 相同的 Source）。

6. 扩展资料

（1）实验 1：Flume 安装

① 下载 apache-flume-1.9.0-bin.tar.gz 安装包并解压，如图 4-25 所示。解压后会出现蓝色的文件，如图 4-26 所示。

② 切换到 flume 目录并复制配置文件/conf/flume-env.sh.template，命名为“flume-env.sh”，如图 4-27 和图 4-28 所示。

```
[root@hservermain software]# tar -zxvf /install/apache-flume-1.9.0-bin.tar.gz
```

图 4-25 解压安装文件

```
[root@hservermain software]# ls
apache-flume-1.9.0-bin  apache-storm-1.2.2
```

图 4-26 查看安装结果

```
[root@hservermain apache-flume-1.9.0-bin]# cp conf/flume-env.sh.template conf/flume-env.sh
```

图 4-27 创建配置文件

```
[root@hservermain apache-flume-1.9.0-bin]# ls ./conf/
flume-conf.properties.template  flume-env.ps1.template  flume-env.sh  flume-env.sh.template  log4j.properties
[root@hservermain apache-flume-1.9.0-bin]#
```

图 4-28 查看配置文件列表

③ 编辑 flume-env.sh,修改 JAVA_HOME 为本机 Java 安装目录,如图 4-29 所示。

```
# Licensed to the Apache Software Foundation (ASF) under one
# or more contributor license agreements.  See the NOTICE file
# distributed with this work for additional information
# regarding copyright ownership.  The ASF licenses this file
# to you under the Apache License, Version 2.0 (the
# "License"); you may not use this file except in compliance
# with the License.  You may obtain a copy of the License at
#
#     http://www.apache.org/licenses/LICENSE-2.0
#
# Unless required by applicable law or agreed to in writing, software
# distributed under the License is distributed on an "AS IS" BASIS,
# WITHOUT WARRANTIES OR CONDITIONS OF ANY KIND, either express or implied.
# See the License for the specific language governing permissions and
# limitations under the License.

# If this file is placed at FLUME_CONF_DIR/flume-env.sh, it will be sourced
# during Flume startup.

# Enviroment variables can be set here.

export JAVA_HOME=/software/java

# Give Flume more memory and pre-allocate, enable remote monitoring via JMX
# export JAVA_OPTS="-Xms100m -Xmx2000m -Dcom.sun.management.jmxremote"

# Let Flume write raw event data and configuration information to its log files for debugging
# purposes. Enabling these flags is not recommended in production,
# as it may result in logging sensitive user information or encryption secrets.
# export JAVA_OPTS="$JAVA_OPTS -Dorg.apache.flume.log.rawdata=true -Dorg.apache.flume.log.printconfig=true "

# Note that the Flume conf directory is always included in the classpath.
#FLUME_CLASSPATH=""

-- INSERT --
```

图 4-29 修改配置文件

④ 验证是否安装成功:输入“bin/flume-ng version”,出现版本信息等则表示安装成功,如图 4-30 所示。

```
[root@hservermain apache-flume-1.9.0-bin]# bin/flume-ng version
Flume 1.9.0
Source code repository: https://git-wip-us.apache.org/repos/asf/flume.git
Revision: d4fcab4f501d41597bc616921329a4339f73585e
Compiled by fszabo on Mon Dec 17 20:45:25 CET 2018
From source with checksum 35db629a3bda49d23e9b3690c80737f9
[root@hservermain apache-flume-1.9.0-bin]#
```

图 4-30 查看 Flume 版本信息

(2) 实验 2:Flume 获取数据并导入 HDFS

① 创建 agent 配置文件,切换到 flume 目录输入“vi conf/spool. conf”编辑文件,如图 4-31 所示。这里源使用 Spool,通道使用内存通道,水槽使用 HDFS。Spool 监测配置的目录下新增的

文件，并将文件中的数据读取出来，然后写到 Channel，写入后，标记该文件已完成或者删除文件。

```
[root@hservermain apache-flume-1.9.0-bin]# vi conf/spool.conf
```

图 4-31 查看配置文件

spool. conf 文件的内容如图 4-32 所示。

```
a1.sources = r1
a1.sinks = k1
a1.channels = c1

# Describe/configure the source
a1.sources.r1.type = spooldir
a1.sources.r1.channels = c1
a1.sources.r1.spoolDir = /software/apache-flume-1.9.0-bin/logs
a1.sources.r1.fileHeader = true

# Describe the sink
a1.sinks.k1.type = hdfs
a1.sinks.k1.hdfs.path = hdfs://hservermain:9000/flume/%y-%m-%d/%H%M
a1.sinks.k1.hdfs.useLocalTimeStamp = true
a1.sinks.k1.hdfs.round = true
a1.sinks.k1.hdfs.roundValue = 10
a1.sinks.k1.hdfs.roundUnit = minute

# Use a channel which buffers events in memory
a1.channels.c1.type = memory
a1.channels.c1.capacity = 1000
a1.channels.c1.transactionCapacity = 100

# Bind the source and sink to the channel
a1.sources.r1.channels = c1
a1.sinks.k1.channel = c1
~
~
~
~
~
~
~
~
-- INSERT --
```

图 4-32 配置文件内容

其中主要有两处需要说明。a1. sources. r1. spoolDir 的参数代表 Spool 源实时监控的目录，当目录中有新的文件生成时，自动搜集；a1. sinks. k1. hdfs. path 的参数代表 Sink 从通道读取数据上传到 HDFS 的目录，“hservermain:9000”是 HDFS 的主机名与端口，在 Hadoop 相关配置文件中配置，需要注意主机名不能写成“localhost”或“127. 0. 0. 1”，因为 9000 端口默认是拒绝 127. 0. 0. 1 访问的，这一点可以通过启动 Hadoop 后使用“telnet localhost 9000”连接端口尝试(会被拒绝访问，改成 IP 或者主机名可连接)，“%y-%m-%d”以及“%H%M”为时间控制输出的格式。

② 创建数据收集源目录/opt/flume/logs，如图 4-33 和图 4-34 所示。

```
[root@hservermain apache-flume-1.9.0-bin]# mkdir /software/apache-flume-1.9.0-bin/logs
[root@hservermain apache-flume-1.9.0-bin]#
```

图 4-33 创建数据收集源目录

```
[root@hservermain apache-flume-1.9.0-bin]# ls
bin  CHANGELOG  conf  DEVNOTES  doap_Flume.rdf  docs  lib  LICENSE  logs  NOTICE  README.md  RELEASE-NOTES  tools
[root@hservermain apache-flume-1.9.0-bin]#
```

图 4-34 查看创建后的目录

③ 启动 Hadoop：使用命令“start-all. sh”(可以不启动 YARN，直接使用命令“start-dfs. sh”即可)，如图 4-35 所示，如果没有配置环境变量，需要加上“start-all. sh”的路径。

```
[root@hservermain software]# start-all.sh
```

图 4-35 启动 Hadoop

启动完成后使用 jps 命令查看进程，若出现图 4-36 所示信息则代表启动成功。

```
[root@hservermain hadoop-3.2.0]# jps
4848 Jps
3890 NameNode
4610 NodeManager
4003 DataNode
4182 SecondaryNameNode
1258 QuorumPeerMain
4494 ResourceManager
```

图 4-36　进程列表

④ 启动 Flume，如图 4-37 所示。

```
[root@hservermain apache-flume-1.9.0-bin]# bin/flume-ng agent -n a1 -c conf -f conf/spool.conf -Dflum
e.root.logger=INFO,console
```

图 4-37　启动 Flume

若 Channel、Sink、Source 都启动成功，则会出现图 4-38 所示的信息。

```
l counters:{} } }} channels:{c1=org.apache.flume.channel.MemoryChannel{name: c1}} }
2019-09-26 05:18:15,681 (conf-file-poller-0) [INFO - org.apache.flume.node.Application.startAllCompon
ents(Application.java:169)] Starting Channel c1
2019-09-26 05:18:15,687 (conf-file-poller-0) [INFO - org.apache.flume.node.Application.startAllCompon
ents(Application.java:184)] Waiting for channel: c1 to start. Sleeping for 500 ms
2019-09-26 05:18:16,019 (lifecycleSupervisor-1-0) [INFO - org.apache.flume.instrumentation.MonitoredC
ounterGroup.register(MonitoredCounterGroup.java:119)] Monitored counter group for type: CHANNEL, name
: c1: Successfully registered new MBean.
2019-09-26 05:18:16,020 (lifecycleSupervisor-1-0) [INFO - org.apache.flume.instrumentation.MonitoredC
ounterGroup.start(MonitoredCounterGroup.java:95)] Component type: CHANNEL, name: c1 started
2019-09-26 05:18:16,188 (conf-file-poller-0) [INFO - org.apache.flume.node.Application.startAllCompon
ents(Application.java:196)] Starting Sink k1
2019-09-26 05:18:16,191 (conf-file-poller-0) [INFO - org.apache.flume.node.Application.startAllCompon
ents(Application.java:207)] Starting Source r1
2019-09-26 05:18:16,200 (lifecycleSupervisor-1-4) [INFO - org.apache.flume.source.SpoolDirectorySourc
e.start(SpoolDirectorySource.java:85)] SpoolDirectorySource source starting with directory: /software
/apache-flume-1.9.0-bin/logs
2019-09-26 05:18:16,244 (lifecycleSupervisor-1-1) [INFO - org.apache.flume.instrumentation.MonitoredC
ounterGroup.register(MonitoredCounterGroup.java:119)] Monitored counter group for type: SINK, name: k
1: Successfully registered new MBean.
2019-09-26 05:18:16,245 (lifecycleSupervisor-1-1) [INFO - org.apache.flume.instrumentation.MonitoredC
ounterGroup.start(MonitoredCounterGroup.java:95)] Component type: SINK, name: k1 started
2019-09-26 05:18:16,456 (lifecycleSupervisor-1-4) [INFO - org.apache.flume.instrumentation.MonitoredC
ounterGroup.register(MonitoredCounterGroup.java:119)] Monitored counter group for type: SOURCE, name:
 r1: Successfully registered new MBean.
2019-09-26 05:18:16,457 (lifecycleSupervisor-1-4) [INFO - org.apache.flume.instrumentation.MonitoredC
ounterGroup.start(MonitoredCounterGroup.java:95)] Component type: SOURCE, name: r1 started
```

图 4-38　Flume 启动结果

此处不要在启动进程时加上“&”让其在后台运行，因为接下来实时收集数据时需要观察这个进程的实时日志信息。

⑤ 在虚拟终端或 Xshell 中对虚拟机另外开启一个会话，如图 4-39 所示。

图 4-39　对虚拟机另外开启会话

⑥ 在新的会话中查看 logs 中无文件，新建一个 test. log 文件，如图 4-40 所示，在 test. log 文件中编辑一些内容并将文件移动到 logs 目录下，如图 4-41 和图 4-42 所示。

```
[root@hservermain apache-flume-1.9.0-bin]# ls logs/
[root@hservermain apache-flume-1.9.0-bin]# vi test.log
```

图 4-40 创建测试文件

```
Apache Flume is a distributed,reliable,and available
system for efficiently collection,aggregating and moving
large amounts of log data from many different sources to
a centralized data store.
~
~
~
~
~
~
~
~
~
~
~
~
~
~
~
~
~
~
~
~
~
~
~
~
"test.log" 4L, 193C
```

图 4-41 编辑测试文件

```
[root@hservermain apache-flume-1.9.0-bin]# cp test.log logs/
[root@hservermain apache-flume-1.9.0-bin]#
```

图 4-42 移动文件

⑦ 可以看到启动 Flume 进程的终端实时地输出了一些日志信息，如图 4-43 所示。

```
2019-09-26 05:18:16,457 (lifecycleSupervisor-1-4) [INFO - org.apache.flume.instrumentation.MonitoredC
ounterGroup.start(MonitoredCounterGroup.java:95)] Component type: SOURCE, name: r1 started
2019-09-26 05:19:45,778 (pool-3-thread-1) [INFO - org.apache.flume.client.avro.ReliableSpoolingFileEv
entReader.readEvents(ReliableSpoolingFileEventReader.java:384)] Last read took us just up to a file b
oundary. Rolling to the next file, if there is one.
2019-09-26 05:19:45,779 (pool-3-thread-1) [INFO - org.apache.flume.client.avro.ReliableSpoolingFileEv
entReader.rollCurrentFile(ReliableSpoolingFileEventReader.java:497)] Preparing to move file /software
/apache-flume-1.9.0-bin/logs/test.log to /software/apache-flume-1.9.0-bin/logs/test.log.COMPLETED
2019-09-26 05:19:50,333 (SinkRunner-PollingRunner-DefaultSinkProcessor) [INFO - org.apache.flume.sink
.hdfs.HDFSSequenceFile.configure(HDFSSequenceFile.java:63)] writeFormat = Writable, UseRawLocalFileSy
stem = false
2019-09-26 05:19:51,256 (SinkRunner-PollingRunner-DefaultSinkProcessor) [INFO - org.apache.flume.sink
.hdfs.BucketWriter.open(BucketWriter.java:246)] Creating hdfs://hservermain:8020/flume/19-09-26/0510/
FlumeData.1569446390328.tmp
2019-09-26 05:20:00,101 (SinkRunner-PollingRunner-DefaultSinkProcessor) [INFO - org.apache.flume.sink
.hdfs.HDFSSequenceFile.configure(HDFSSequenceFile.java:63)] writeFormat = Writable, UseRawLocalFileSy
stem = false
2019-09-26 05:20:00,230 (SinkRunner-PollingRunner-DefaultSinkProcessor) [INFO - org.apache.flume.sink
.hdfs.BucketWriter.open(BucketWriter.java:246)] Creating hdfs://hservermain:8020/flume/19-09-26/0520/
FlumeData.1569446400102.tmp
2019-09-26 05:20:30,046 (hdfs-k1-roll-timer-0) [INFO - org.apache.flume.sink.hdfs.HDFSEventSink$1.run
(HDFSEventSink.java:393)] Writer callback called.
2019-09-26 05:20:30,047 (hdfs-k1-roll-timer-0) [INFO - org.apache.flume.sink.hdfs.BucketWriter.doClos
e(BucketWriter.java:438)] Closing hdfs://hservermain:8020/flume/19-09-26/0510/FlumeData.1569446390328
.tmp
2019-09-26 05:20:30,138 (hdfs-k1-call-runner-9) [INFO - org.apache.flume.sink.hdfs.BucketWriter$7.cal
l(BucketWriter.java:681)] Renaming hdfs://hservermain:8020/flume/19-09-26/0510/FlumeData.156944639032
```

图 4-43 查看终端输出信息

在浏览器中输入 192.168.159.130:50070,在 Namenode 页面单击"Utilities→Browse the file system"可以看到 HDFS 的相关存储文件,如图 4-44 所示。

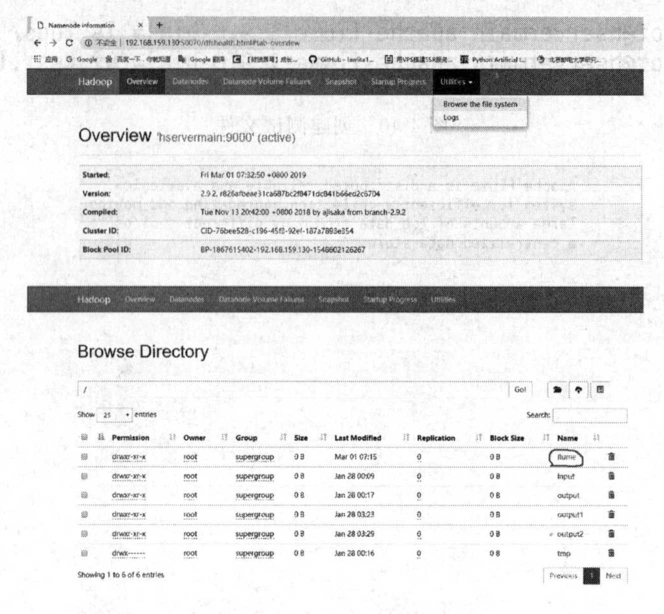

图 4-44　管理页 HDFS 信息

可以看到在 HDFS 中的"/flume/19-02-28/1540"有一个文件"FlumeData.1551397475841",如图 4-45 所示,其中的数字是时间戳,目录结构为之前在配置文件中设置的目录,可以看到文件创建时间为 2019 年 2 月 28 日的 15:40,这个时间是 UTC 时间,和北京时间相差 12 小时。

图 4-45　HDFS 上文件详情

⑧ 单击文件,可以看到文件的一些信息。单击"Download",下载到本地,如图 4-46 所示。

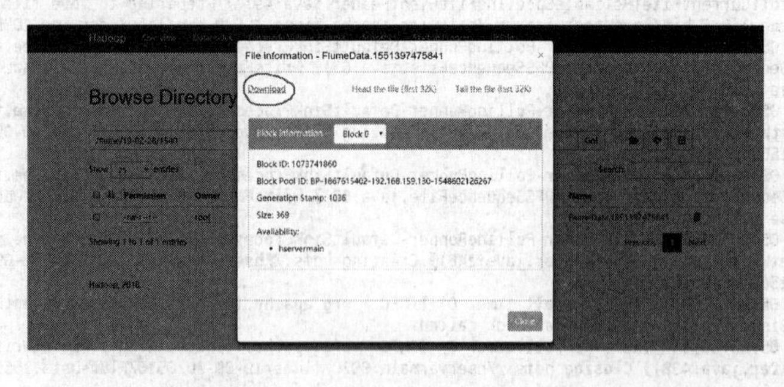

图 4-46　下载 HDFS 上的文件

7. 参考答案

实验作业①和②的答案见本节扩展资料部分。实验作业③可参考其他相关资料完成。

4.3　Kafka 安装及其基本操作

1. 实验目的

学会在计算机上安装消息系统 Kafka 并掌握消息生产到消费的整个生命周期。

2. 实验要求

① Java 8。

② Linux、MacOSX 或其他类似于 Unix 的操作系统(不支持 Windows)。

③ 8 GB 内存。

④ 2 个 vCPU。

⑤ ZooKeeper 分布式协调服务。

3. 预备知识

Apache Kafka 是一个分布式的消息系统,它作为一个数据管道实现了不同系统间的实时通信,在离线、近似实时以及实时系统中都可以应用。Kafka 依赖 ZooKeeper 管理和协调其每个 broker 节点。

4. 实验内容[25]

(1) 实验 1:Kafka 安装

① 下载 kafka_2.12-2.1.1.tgz 安装包。

② 终端运行 tar 命令解压安装包。

```
tar -zxvf /install/kafka_2.12-2.1.1.tgz
```

③ 启动 ZooKeeper 服务。

```
bin/zookeeper-server-start.sh config/zookeeper.properties &
```

④ 启动 Kafka 服务。

```
bin/kafka-server-start.sh -daemon config/server.properties
```

⑤ 查看进程启动状态。

```
jps
```

(2) 实验 2:Kafka 消息发送与接收

① 创建名为 test 的 Topic 并查看。

创建:

```
bin/kafka-topics.sh --create --zookeeper localhost:2181 --replication-factor 1 --partitions 1  --topic test
```

查看:

```
bin/kafka-topics.sh --list --zookeeper localhost:2181
```

② 生产者发送消息,消费者实时接收。

生产者端:

```
bin/kafka-console-producer.sh --broker-list localhost:9092 --topic test
```

消费者端:

```
bin/kafka-console-consumer.sh --bootstrap-server localhost:9092 --topic test --from-beginning
```

5. 实验作业

① 下载 Kafka 安装包并在虚拟机上安装。

② 使用 Kafka 实时发送、接收消息。

6. 扩展资料

(1) 实验 1:Kafka 安装

① 下载 apache-flume-1.9.0-bin.tar.gz 安装包并解压,如图 4-47 所示。

```
[root@hservermain software]# tar -zxvf /install/kafka_2.12-2.1.1.tgz
```

图 4-47 解压安装文件

解压后会出现蓝色的文件,如图 4-48 所示。

```
[root@hservermain software]# ls
apache-storm-1.2.2  java  kafka_2.12-2.1.1  zookeeper-3.4.9
```

图 4-48 查看安装结果

② Kafka 中集成了 ZooKeeper 服务,这里直接使用集成的服务,当然也可以使用单独下载的 ZooKeeper 开启服务。查看 ZooKeeper 服务的配置文件,可以看到默认的端口是 2181,如图 4-49 和图 4-50 所示。

```
[root@hservermain kafka_2.12-2.1.1]# vi config/zookeeper.properties
```

图 4-49 查看 ZooKeeper 配置文件

```
# Licensed to the Apache Software Foundation (ASF) under one or more
# contributor license agreements.  See the NOTICE file distributed with
# this work for additional information regarding copyright ownership.
# The ASF licenses this file to You under the Apache License, Version 2.0
# (the "License"); you may not use this file except in compliance with
# the License.  You may obtain a copy of the License at
#
#    http://www.apache.org/licenses/LICENSE-2.0
#
# Unless required by applicable law or agreed to in writing, software
# distributed under the License is distributed on an "AS IS" BASIS,
# WITHOUT WARRANTIES OR CONDITIONS OF ANY KIND, either express or implied.
# See the License for the specific language governing permissions and
# limitations under the License.
# the directory where the snapshot is stored.
dataDir=/tmp/zookeeper
# the port at which the clients will connect
clientPort=2181
# disable the per-ip limit on the number of connections since this is a non-production config
maxClientCnxns=0
```

图 4-50 查看配置文件信息

③ 启动 ZooKeeper 进程,如图 4-51 所示,输入以下命令。

```
bin/zookeeper-server-start.sh config/zookeeper.properties
```

```
[root@hservermain kafka_2.12-2.1.1]# bin/zookeeper-server-start.sh config/zookeeper.properties &
[1] 12339
```

图 4-51 启动 ZooKeeper

使用 jps 命令查看进程是否开启,如图 4-52 所示。

```
[root@hservermain kafka_2.12-2.1.1]# jps
12640 Jps
12339 QuorumPeerMain
```

图 4-52 进程列表

此处之所以不放到后台启动 ZooKeeper，是因为其管理着 Kafka，当对 Kafka 进行一些操作时会实时输出相关日志信息，会使操作结果夹杂着日志信息，导致输出很混乱。

④ 后台启动 Kafka 服务。重新打开一个虚拟机终端窗口，查看 Kafka 服务配置文件 server.properties，如图 4-53 和图 4-54 所示，由于是单节点启动，因此不用修改配置信息，需要注意服务默认端口是 9092，与 ZooKeeper 连接的默认端口是 2181(与上面的 ZooKeeper 配置需要对应)，如图 4-55 和图 4-56 所示。实际的集群生产环境下，需要修改相关配置。

```
[root@hservermain kafka_2.12-2.1.1]# vi config/server.properties
```

图 4-53 查看 Kafka 配置文件

```
# Licensed to the Apache Software Foundation (ASF) under one or more
# contributor license agreements.  See the NOTICE file distributed with
# this work for additional information regarding copyright ownership.
# The ASF licenses this file to You under the Apache License, Version 2.0
# (the "License"); you may not use this file except in compliance with
# the License.  You may obtain a copy of the License at
#
#    http://www.apache.org/licenses/LICENSE-2.0
#
# Unless required by applicable law or agreed to in writing, software
# distributed under the License is distributed on an "AS IS" BASIS,
# WITHOUT WARRANTIES OR CONDITIONS OF ANY KIND, either express or implied.
# See the License for the specific language governing permissions and
# limitations under the License.

# see kafka.server.KafkaConfig for additional details and defaults

############################# Server Basics #############################

# The id of the broker. This must be set to a unique integer for each broker.
broker.id=0

############################# Socket Server Settings #############################

# The address the socket server listens on. It will get the value returned from
# java.net.InetAddress.getCanonicalHostName() if not configured.
#   FORMAT:
#     listeners = listener_name://host_name:port
"config/server.properties" [noeol] 136L, 6851C
```

图 4-54 Kafka 配置文件(一)

```
############################# Socket Server Settings #############################

# The address the socket server listens on. It will get the value returned from
# java.net.InetAddress.getCanonicalHostName() if not configured.
#   FORMAT:
#     listeners = listener_name://host_name:port
#   EXAMPLE:
#     listeners = PLAINTEXT://your.host.name:9092
#listeners=PLAINTEXT://:9092
```

图 4-55 Kafka 配置文件(二)

输入命令“bin/kafka-server-start. sh -daemon config/server. properties”后台启动 Kafka 进程，如图 4-57 所示。

```
############################# Zookeeper #############################

# Zookeeper connection string (see zookeeper docs for details).
# This is a comma separated host:port pairs, each corresponding to a zk
# server. e.g. "127.0.0.1:3000,127.0.0.1:3001,127.0.0.1:3002".
# You can also append an optional chroot string to the urls to specify the
# root directory for all kafka znodes.
zookeeper.connect=localhost:2181

# Timeout in ms for connecting to zookeeper
zookeeper.connection.timeout.ms=6000
```

图 4-56 Kafka 配置文件(三)

```
[root@hservermain kafka_2.12-2.1.1]# bin/kafka-server-start.sh -daemon config/server.properties
```

图 4-57 启动 Kafka

使用 jps 命令查看,如图 4-58 所示。

```
[root@hservermain kafka_2.12-2.1.1]# jps
12339 QuorumPeerMain
12922 Kafka
12939 Jps
```

图 4-58 进程列表

(2) 实验 2:Kafka 消息发送与接收

① 创建 Topic 并查看。输入下述命令创建一个名为 test 的 Topic,如图 4-59 所示。

```
bin/kafka-topics.sh --create --zookeeper localhost:2181 --replication-factor 1
--partitions 1  --topic test
```

```
[root@hservermain kafka_2.12-2.1.1]# bin/kafka-topics.sh --create --zookeeper localhost:2181 --replication-
factor 1 --partitions 1  --topic test
[2019-09-26 03:31:18,625] INFO Accepted socket connection from /0:0:0:0:0:0:0:1:47532 (org.apache.zookeeper
.server.NIOServerCnxnFactory)
[2019-09-26 03:31:18,643] INFO Client attempting to establish new session at /0:0:0:0:0:0:0:1:47532 (org.ap
ache.zookeeper.server.ZooKeeperServer)
[2019-09-26 03:31:18,650] INFO Established session 0x1000149496f0001 with negotiated timeout 30000 for clie
nt /0:0:0:0:0:0:0:1:47532 (org.apache.zookeeper.server.ZooKeeperServer)
[2019-09-26 03:31:20,794] INFO Got user-level KeeperException when processing sessionid:0x1000149496f0001 t
ype:setData cxid:0x4 zxid:0x1f txntype:-1 reqpath:n/a Error Path:/config/topics/test Error:KeeperErrorCode
= NoNode for /config/topics/test (org.apache.zookeeper.server.PrepRequestProcessor)
Created topic "test".
[2019-09-26 03:31:21,344] INFO Processed session termination for sessionid: 0x1000149496f0001 (org.apache.z
ookeeper.server.PrepRequestProcessor)
[2019-09-26 03:31:21,353] INFO Closed socket connection for client /0:0:0:0:0:0:0:1:47532 which had session
id 0x1000149496f0001 (org.apache.zookeeper.server.NIOServerCnxn)
[root@hservermain kafka_2.12-2.1.1]#
```

图 4-59 创建 Topic

输入下述命令查看 Topic,如图 4-60 所示。

```
bin/kafka-topics.sh --list --zookeeper localhost:2181
```

```
[root@hservermain kafka_2.12-2.1.1]# bin/kafka-topics.sh --list --zookeeper localhost:2181
[2019-09-26 03:32:21,273] INFO Accepted socket connection from /0:0:0:0:0:0:0:1:47534 (org.apache.zookeeper
.server.NIOServerCnxnFactory)
[2019-09-26 03:32:21,291] INFO Client attempting to establish new session at /0:0:0:0:0:0:0:1:47534 (org.ap
ache.zookeeper.server.ZooKeeperServer)
[2019-09-26 03:32:21,299] INFO Established session 0x1000149496f0002 with negotiated timeout 30000 for clie
nt /0:0:0:0:0:0:0:1:47534 (org.apache.zookeeper.server.ZooKeeperServer)
test
[2019-09-26 03:32:21,623] INFO Processed session termination for sessionid: 0x1000149496f0002 (org.apache.z
ookeeper.server.PrepRequestProcessor)
[2019-09-26 03:32:21,631] INFO Closed socket connection for client /0:0:0:0:0:0:0:1:47534 which had session
id 0x1000149496f0002 (org.apache.zookeeper.server.NIOServerCnxn)
[root@hservermain kafka_2.12-2.1.1]#
```

图 4-60 查看 Topic

② 生产者发送消息，消费者实时接收。重新打开一个虚拟机终端窗口，暂且称作“消费者端”，当前窗口称作“生产者端”。

在生产者端输入下述命令，如图 4-61 所示。

```
bin/kafka-console-producer.sh --broker-list localhost:9092 --topic test
```

```
[root@hservermain kafka_2.12-2.1.1]# bin/kafka-console-producer.sh --broker-list localhost:9092 --topic test
```

图 4-61　创建生产者

在消费者端切换到 kafka 目录并输入下述命令，如图 4-62 所示。

```
bin/kafka-console-consumer.sh --bootstrap-server localhost:9092 --topic test
--from-beginning
```

```
[root@hservermain kafka_2.12-2.1.1]# bin/kafka-console-consumer.sh --bootstrap-server localhost:9092
--topic test --from-beginning
```

图 4-62　创建消费者

在生产者端输出一段消息并按下“Enter”键，如图 4-63 所示。

```
[root@hservermain kafka_2.12-2.1.1]# bin/kafka-console-producer.sh --broker-list localhost:9092 --top
ic test
>This is my first kafka test
>
```

图 4-63　生产者发送消息

可以看到消费者端实时地接收到了消息，如图 4-64 所示。

```
[root@hservermain kafka_2.12-2.1.1]# bin/kafka-console-consumer.sh --bootstrap-server localhost:9092
--topic test --from-beginning
This
This is my first test
This is my first kafka test
```

图 4-64　消费者接收消息

7. 参考答案

实验作业的答案见本节扩展资料部分。

4.4　Spark Streaming 安装及其基本操作

1. 实验目的

学会在计算机上安装实时处理框架 Spark，并学会使用其重要组件 Spark Streaming。

2. 实验要求

① Java 8。

② Scala 2.11.12。

③ Linux、MacOSX 或其他类似于 Unix 的操作系统（不支持 Windows）。

④ 8 GB 内存。

⑤ 2 个 vCPU。

⑥ SBT。

⑦ IDEA。

3. 预备知识

Apache Spark 的产生是为了解决 MapReduce 的计算缓慢问题，其初衷是设计一个可以解决所有类型的计算的引擎，包括离线批处理、交互式查询、实时流式处理、图计算以及迭代计算等。Spark Streaming 是 Spark 最重要的组件之一，其功能与 Storm 相同，都是进行流式计算的。

SBT(Simple Build Tool)是类似于 Ant、Maven 的构建工具，是 Scala 的标准构建工具。

拓展阅读

Spark Streaming 介绍

4. 实验内容

(1) 实验 1：Spark 安装[26]

① 下载 spark-2.4.0-bin-hadoop2.7.tgz 安装包。

② 终端运行 tar 命令解压安装包。

```
tar -zxvf /install/spark-2.4.0-bin-hadoop2.7.tgz
```

③ 修改配置文件。复制 conf 下的 spark-env.sh.template，更名为 spark-env.sh。

```
cp conf/spark-env.sh.template conf/spark-env.sh
```

使用命令“vi conf/spark-env.sh”编辑其中的内容。

```
export JAVA_HOME = /software/java
export SPARK_MASTER_IP = hservermain
export SPARK_MASTER_PORT = 7077
```

复制 conf 下的 slaves.template，更名为 slaves。

```
cp conf/slaves.template conf/slaves
```

使用命令“vi conf/slaves”编辑其中的内容(由于是单节点，因此只需要从节点改成主机名即可)。

```
hservermain
```

④ 启动并验证是否成功。

```
./sbin/start.all
jps
```

在浏览器中输入“主机 IP：端口”查看 Spark 运行状态。

```
192.168.159.130:8080
```

(2) 实验 2：Spark 的三种运行方式

① 开启新的终端窗口(2 号)，开启 9999 端口以实时写入。

```
nc -lk 9999
```

② 在 1 号终端使用“run-example”命令运行示例程序。

```
bin/run-example streaming.NetworkWordCount localhost 9999
```

在 2 号终端实时输入消息并观察 1 号终端程序的实时输出情况。

③ 在 1 号终端使用“spark-shell”命令交互式输入程序。

```
bin/spark-shell
```

等待一段时间后进入 Shell，交互式(逐句)输入以下代码。

```
import org.apache.spark.SparkConf
import org.apache.spark.streaming.{ Seconds, StreamingContext }
val ssc = new StreamingContext(sc, Seconds(5))
val lines = ssc.socketTextStream("localhost", 9999)
val words = lines.flatMap(_.split(" "))
val wordCounts = words.map(x => (x, 1)).reduceByKey(_ + _)
wordCounts.print()
ssc.start()
```

在 2 号终端实时输入消息，并观察 1 号终端程序的实时输出情况。

④ 本地开发 SBT 项目并打包。

⑤ 将打包程序上传到服务器端。

⑥ 在 1 号终端使用“spark-submit”命令运行程序。

```
./bin/spark-submit --class WordCount ../sparkStreamingTest.jar
```

在 2 号终端实时写入消息，并观察 1 号终端程序的实时输出情况。

⑦ 观察浏览器端任务的运行情况。

```
192.168.159.130:8080
```

5. 实验作业

① 下载 Spark 安装包并在虚拟机上安装。

② 在服务器端使用 Spark 的三种程序运行方式运行 Spark Streaming 程序。

6. 扩展资料

(1) 实验 1：Spark 安装

① 下载 spark-2.4.0-bin-hadoop2.7.tgz 安装包并解压，这是集成了 Hadoop 2.7 环境的 Spark 安装包，如果已经安装了 Hadoop 或后期用不到 Hadoop 生态组件则可以选择未集成 Hadoop 环境的安装包：spark-2.4.0-bin-without-hadoop-scala-2.12.tgz，如图 4-65 所示。

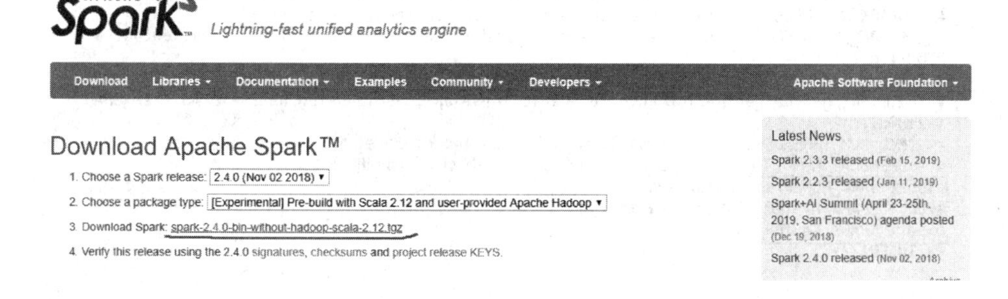

图 4-65 Spark 下载

本次实验用不到 Hadoop 生态组件，因此选择上述两种安装包均可，以前者为例进行安装，如图 4-66 所示。

```
[root@hservermain software]# tar -zxvf /install/spark-2.4.0-bin-hadoop2.7.tgz
```

图 4-66　解压安装文件

解压后会出现蓝色的文件，如图 4-67 所示。

```
[root@hservermain software]# ls
apache-flume-1.9.0-bin  java
apache-maven-3.6.2      kafka_2.12-2.1.1
apache-storm-1.2.2      spark-2.4.0-bin-hadoop2.7
flink-1.7.2             wiki-edits
hadoop-2.9.2            wiki-edits-0.1.jar
```

图 4-67　查看安装结果

② 复制一份 conf 目录下的配置文件 spark-env. sh. template，更改文件名为 spark-env. sh 并编辑其内容，如图 4-68 和图 4-69 所示。

```
[root@hservermain spark-2.4.0-bin-hadoop2.7]# cp conf/spark-env.sh.template conf/spark-en
v.sh
```

图 4-68　创建配置文件

```
[root@hservermain spark-2.4.0-bin-hadoop2.7]# vi conf/spark-env.sh
```

图 4-69　查看 Spark 配置文件

spark-env. sh 中需要配置 Java、Scala 以及 Spark 的相关参数信息，Spark 默认的端口是 7077，配置内容如图 4-70 所示。

```
# - SPARK_DAEMON_MEMORY, to allocate to the master, worker and history server themselves
(default: 1g).
# - SPARK_HISTORY_OPTS, to set config properties only for the history server (e.g. "-Dx=y
")
# - SPARK_SHUFFLE_OPTS, to set config properties only for the external shuffle service (e
.g. "-Dx=y")
# - SPARK_DAEMON_JAVA_OPTS, to set config properties for all daemons (e.g. "-Dx=y")
# - SPARK_DAEMON_CLASSPATH, to set the classpath for all daemons
# - SPARK_PUBLIC_DNS, to set the public dns name of the master or workers

# Generic options for the daemons used in the standalone deploy mode
# - SPARK_CONF_DIR      Alternate conf dir. (Default: ${SPARK_HOME}/conf)
# - SPARK_LOG_DIR       Where log files are stored.  (Default: ${SPARK_HOME}/logs)
# - SPARK_PID_DIR       Where the pid file is stored. (Default: /tmp)
# - SPARK_IDENT_STRING  A string representing this instance of spark. (Default: $USER)
# - SPARK_NICENESS      The scheduling priority for daemons. (Default: 0)
# - SPARK_NO_DAEMONIZE  Run the proposed command in the foreground. It will not output a
PID file.
# Options for native BLAS, like Intel MKL, OpenBLAS, and so on.
# You might get better performance to enable these options if using native BLAS (see SPAR
K-21305).
# - MKL_NUM_THREADS=1        Disable multi-threading of Intel MKL
# - OPENBLAS_NUM_THREADS=1   Disable multi-threading of OpenBLAS
export JAVA_HOME=/software/java
export SPARK_MASTER_IP=hservermain
export SPARK_MASTER_PORT=7077
-- INSERT --
```

图 4-70　编辑配置文件

复制一份 conf 目录下的配置文件 slaves. template，更改文件名为 slaves 并编辑其内容，如图 4-71、图 4-72 和图 4-73 所示。

```
[root@hservermain spark-2.4.0-bin-hadoop2.7]# cp conf/slaves.template conf/slaves
```

图 4-71 创建从节点配置文件

```
[root@hservermain spark-2.4.0-bin-hadoop2.7]# vi conf/slaves
```

图 4-72 查看从节点配置文件

```
#
# Licensed to the Apache Software Foundation (ASF) under one or more
# contributor license agreements.  See the NOTICE file distributed with
# this work for additional information regarding copyright ownership.
# The ASF licenses this file to You under the Apache License, Version 2.0
# (the "License"); you may not use this file except in compliance with
# the License.  You may obtain a copy of the License at
#
#    http://www.apache.org/licenses/LICENSE-2.0
#
# Unless required by applicable law or agreed to in writing, software
# distributed under the License is distributed on an "AS IS" BASIS,
# WITHOUT WARRANTIES OR CONDITIONS OF ANY KIND, either express or implied.
# See the License for the specific language governing permissions and
# limitations under the License.
#

# A Spark Worker will be started on each of the machines listed below.
hservermain
~
~
~
~
~
~
-- INSERT --
```

图 4-73 编辑从节点配置文件

③ 启动 Spark 进程并查看。使用命令“./sbin/start. all”启动 Spark 主节点和从节点，由于是单机启动，因此主从节点都运行在一台主机上，如图 4-74 所示。

```
[root@hservermain spark-2.4.0-bin-hadoop2.7]# ./sbin/start-all.sh
starting org.apache.spark.deploy.master.Master, logging to /software/spark-2.4.0-bin-hado
op2.7/logs/spark-root-org.apache.spark.deploy.master.Master-1-hservermain.out
hservermain: starting org.apache.spark.deploy.worker.Worker, logging to /software/spark-2
.4.0-bin-hadoop2.7/logs/spark-root-org.apache.spark.deploy.worker.Worker-1-hservermain.ou
t
[root@hservermain spark-2.4.0-bin-hadoop2.7]# jps
1123 Master
1177 Worker
1209 Jps
[root@hservermain spark-2.4.0-bin-hadoop2.7]#
```

图 4-74 启动 Spark 主从节点并查看进程

④ 在浏览器中输入“主机 IP:端口(8080)”查看 Spark 节点的运行状态，如图 4-75 所示。

(2) 实验 2:Spark 的三种运行方式[27]

① Spark 有三种运行程序的方式，分别是 spark-submit、spark-shell 以及运行其示例程序的 run-example，三个命令都在 bin 目录下，如图 4-76 所示。

图 4-75　Spark 管理页面

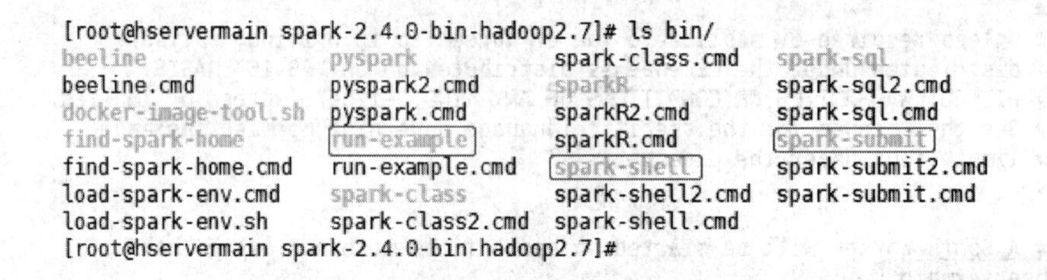

图 4-76　Spark 程序运行文件

② 在 spark 主目录下的 examples/src/main 下有 Java、Python、Scala 和 R 语言的入门示例程序。以 Scala 为例，其关于 Spark Streaming 的示例程序在 examples/src/main/scala/org/apache/spark/examples/streaming 中。下面以实时统计单词个数的入门案例 NetworkWordCount 为例，使用 run-example 方式运行。

在当前终端(1 号)输入“nc -lk 9999”对 9999 端口进行写入操作(如果没有 nc 命令，则使用 yum 安装)，如图 4-77 所示。

```
[root@hservermain spark-2.4.0-bin-hadoop2.7]# nc -lk 9999
```

图 4-77　开启端口

重新打开一个终端窗口(2 号)，切换到 spark 目录，输入以下命令，如图 4-78 所示。

```
bin/run-example streaming.NetworkWordCount localhost 9999
```

```
[root@hservermain spark-2.4.0-bin-hadoop2.7]# bin/run-example streaming.NetworkWordCount
localhost 9999
```

图 4-78　运行 Spark 示例程序

在 1 号终端输入“hello world”，如图 4-79 所示。

```
[root@hservermain spark-2.4.0-bin-hadoop2.7]# nc -lk 9999
hello world
```

图 4-79　向端口进行写操作

可以观察到 2 号终端实时统计出了单词出现频率，这里官网给出的示例是每隔 1 秒统计一次，如图 4-80 所示，因此程序会比较快地闪动。这里需要注意虚拟机必须是 2 核以上 CPU，否则无法正常输出结果。

```
ala:56) finished in 0.055 s
2019-09-28 14:51:16 INFO  DAGScheduler:54 - Job 52 finished: print at NetworkWordCount.sc
ala:56, took 0.067062 s
-------------------------------------------
Time: 1569653476000 ms
-------------------------------------------
(hello,1)
(world,1)

2019-09-28 14:51:16 INFO  JobScheduler:54 - Finished job streaming job 1569653476000 ms.0
 from job set of time 1569653476000 ms
2019-09-28 14:51:16 INFO  JobScheduler:54 - Total delay: 0.328 s for time 1569653476000 m
s (execution: 0.266 s)
2019-09-28 14:51:16 INFO  ShuffledRDD:54 - Removing RDD 100 from persistence list
2019-09-28 14:51:16 INFO  BlockManager:54 - Removing RDD 100
2019-09-28 14:51:16 INFO  MapPartitionsRDD:54 - Removing RDD 99 from persistence list
2019-09-28 14:51:16 INFO  MapPartitionsRDD:54 - Removing RDD 98 from persistence list
2019-09-28 14:51:16 INFO  BlockRDD:54 - Removing RDD 97 from persistence list
2019-09-28 14:51:16 INFO  SocketInputDStream:54 - Removing blocks of RDD BlockRDD[97] at
socketTextStream at NetworkWordCount.scala:53 of time 1569653476000 ms
2019-09-28 14:51:16 INFO  ReceivedBlockTracker:54 - Deleting batches: 1569653474000 ms
2019-09-28 14:51:16 INFO  InputInfoTracker:54 - remove old batch metadata: 1569653474000
ms
2019-09-28 14:51:16 INFO  BlockManager:54 - Removing RDD 99
2019-09-28 14:51:16 INFO  BlockManager:54 - Removing RDD 98
2019-09-28 14:51:16 INFO  BlockManager:54 - Removing RDD 97
```

图 4-80　Spark 示例程序运行结果

③ 可以通过 spark-shell 方式来编写交互式程序以实现增加实时统计间隔。按“Ctrl＋C”终止 1 号终端程序，输入“bin/spark-shell”，如图 4-81 所示。

```
[root@hservermain spark-2.4.0-bin-hadoop2.7]# bin/spark-shell
```

图 4-81　交互式运行方式

此处可以看到一些重要的信息：spark-shell 的 Web 页面 URL，Spark context 对象' sc '(Spark 编程程序的入口，进入 spark-shell 会自动实例化)，Spark 的内置 Scala 版本(这个非常重要，Spark 程序开发时，开发端与服务器端 Scala 版本信息必须对应，否则服务器在运行程序时会出现找不到 Scala 相关类的错误)，如图 4-82 所示。

```
Spark context Web UI available at http://localhost:4040
Spark context available as 'sc' (master = local[*], app id = local-1569653784519).
Spark session available as 'spark'.
Welcome to
```

```
Using Scala version 2.11.12 (Java HotSpot(TM) 64-Bit Server VM, Java 1.8.0_221)
Type in expressions to have them evaluated.
Type :help for more information.

scala>
```

图 4-82　交互式模式的详细信息

④ 交互式输入以下代码，如图 4-83 所示。

```
import org.apache.spark.SparkConf
import org.apache.spark.streaming.{ Seconds, StreamingContext }
val ssc = new StreamingContext(sc, Seconds(5))
val lines = ssc.socketTextStream("localhost", 9999)
val words = lines.flatMap(_.split(" "))
val wordCounts = words.map(x => (x, 1)).reduceByKey(_ + _)
wordCounts.print()
ssc.start()
```

```
scala> import org.apache.spark.SparkConf
import org.apache.spark.SparkConf

scala> import org.apache.spark.streaming.{ Seconds, StreamingContext }
import org.apache.spark.streaming.{Seconds, StreamingContext}

scala> val ssc = new StreamingContext(sc, Seconds(5))
ssc: org.apache.spark.streaming.StreamingContext = org.apache.spark.streaming.StreamingCo
ntext@35d4fecf

scala> val lines = ssc.socketTextStream("localhost", 9999)
lines: org.apache.spark.streaming.dstream.ReceiverInputDStream[String] = org.apache.spark
.streaming.dstream.SocketInputDStream@6c27be1d

scala> val words = lines.flatMap(_.split(" "))
words: org.apache.spark.streaming.dstream.DStream[String] = org.apache.spark.streaming.ds
tream.FlatMappedDStream@74cd82f1

scala> val wordCounts = words.map(x => (x, 1)).reduceByKey(_ + _)
wordCounts: org.apache.spark.streaming.dstream.DStream[(String, Int)] = org.apache.spark.
streaming.dstream.ShuffledDStream@24c6a471

scala> wordCounts.print()
scala> ssc.start()

[Stage 0:>                                                    (0 + 1) / 1]ssc.await
-------------------------------------------
Time: 1569653955000 ms
```

图 4-83 交互式输入程序

在 2 号终端输入“test the shell”，观察 1 号终端实时输出结果，如图 4-84 所示。

```
[root@hservermain spark-2.4.0-bin-hadoop2.7]# nc -lk 9999
test the shell
```

图 4-84 再次向端口进行写操作

1 号终端内程序每隔 5 秒实时统计一次端口发出消息的单词数量，如图 4-85 所示。

```
-------------------------------------------
Time: 1569654060000 ms
-------------------------------------------
(shell,1)
(test,1)
(the,1)

-------------------------------------------
Time: 1569654065000 ms
-------------------------------------------

[Stage 0:>
```

图 4-85 交互式下程序输出结果

⑤ 打开新的终端窗口(3 号)，使用 jps 命令查看 Java 进程，并使用“kill -9 进程号”命令把运行中的 spark-shell 终止，如图 4-86 所示。

```
[root@hservermain ~]# jps
2213 Jps
1177 Master
1242 Worker
1627 SparkSubmit
[root@hservermain ~]# kill -9 1627
[root@hservermain ~]#
```

图 4-86　查看进程列表并“杀死”Spark 进程

⑥ 生产环境中更多的是使用 spark-submit 命令把开发好的 Spark 程序放到 Spark 集群中运行。开发环境使用 IEDA＋Scala＋SBT，其中 SBT 在安装 IDEA 和 Scala 插件后自动集成，Scala 插件在 IDEA 安装后第一次进入时会提示安装，若忘记安装，则可以通过“File→Settings→Plugins”搜索 Scala 进行安装，如图 4-87 所示。

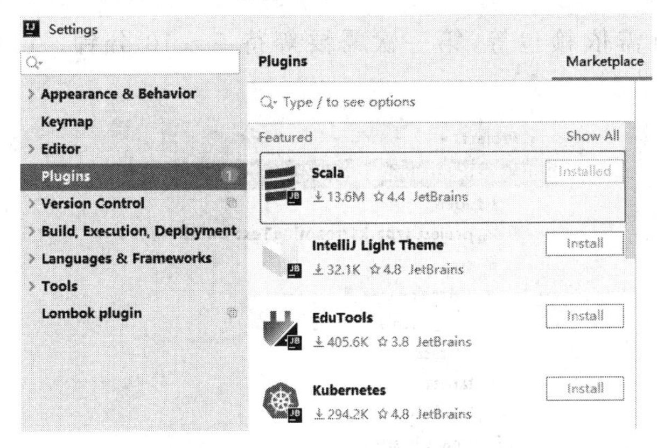

图 4-87　Scala 插件安装

⑦ 创建一个 SBT 项目。选择“File→New→Project”，会出现图 4-88 所示的页面。

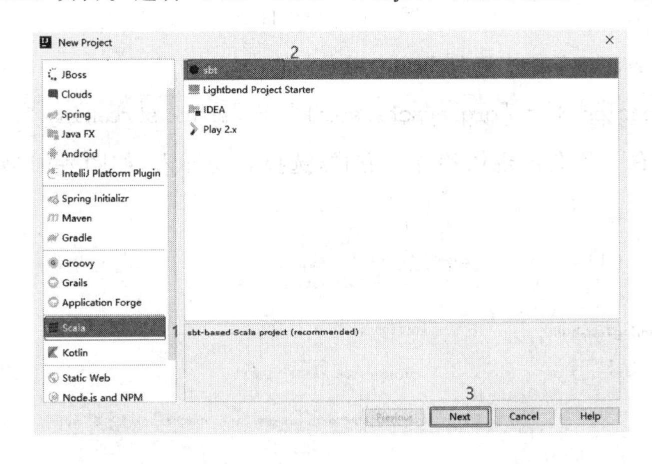

图 4-88　SBT 项目创建

按照图 4-88 所示的步骤执行，进入图 4-89 所示的界面，其中 Scala 版本选择 2.11.12，特别注意此处的 Scala 版本必须与服务器端对应。

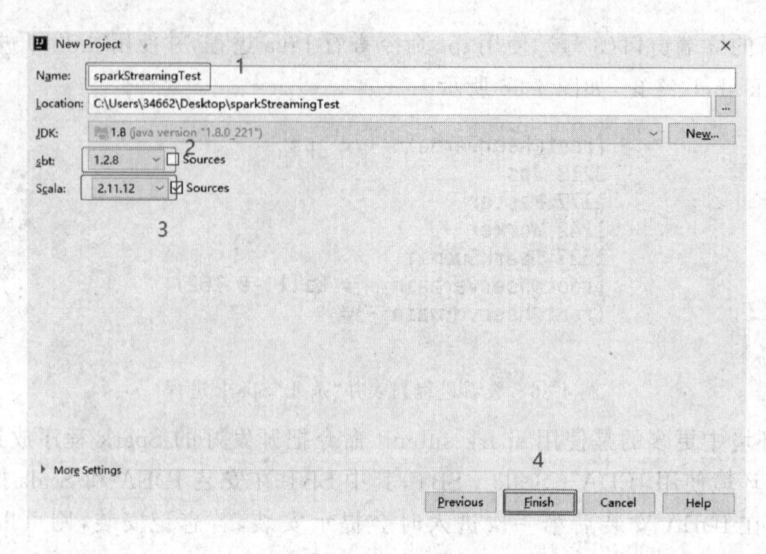

图 4-89　项目环境选择

项目需要加载所需依赖包等，第一次需要等待 5～10 分钟。创建好的项目目录如图 4-90 所示。

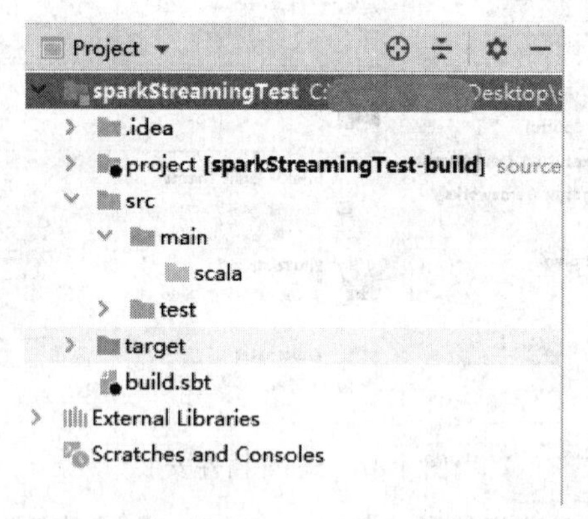

图 4-90　项目目录

⑧ 添加 spark-streaming 依赖到 build. sbt 文件中。

```
libraryDependencies + = "org.apache.spark" % "spark-streaming_2.11" % "2.4.0"
```

添加依赖之后右下角会出现依赖导入方式，选择自动导入，如图 4-91 所示，导入过程需要等待几分钟。

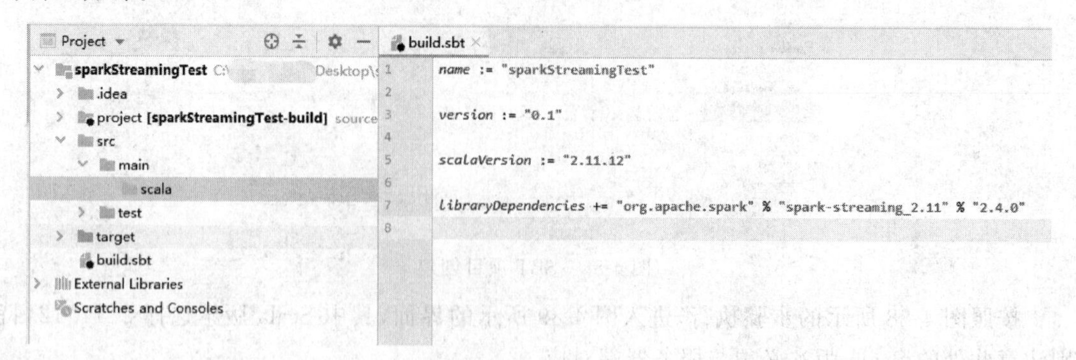

图 4-91　添加 spark-streaming 依赖

⑨ 创建 Scala-Object，如图 4-92 和图 4-93 所示，并输入如下代码，如图 4-94 所示。

```
import org.apache.spark.SparkConf
import org.apache.spark.streaming.{Seconds, StreamingContext}

object WordCount {
  def main(args: Array[String]): Unit = {
    val conf = new SparkConf().setMaster("spark://hservermain:7077")
    .setAppName("NetworkWordCount")
    val ssc = new StreamingContext(conf, Seconds(5))
    val lines = ssc.socketTextStream("localhost", 9999)
    val words = lines.flatMap(_.split(" "))
    val wordCounts = words.map(word => (word, 1)).reduceByKey(_ + _)
    wordCounts.print()
    ssc.start()
    ssc.awaitTermination()
  }
}
```

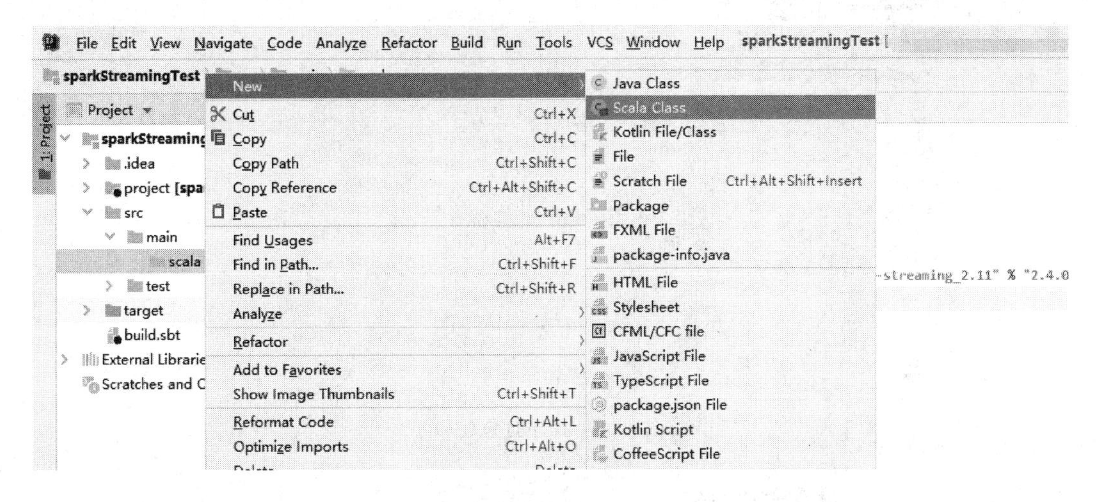

图 4-92 创建 Scala 类

Create New Scala Class

WordCount

Class
Case Class
Object
Case Object
Trait

图 4-93 选择对象类型

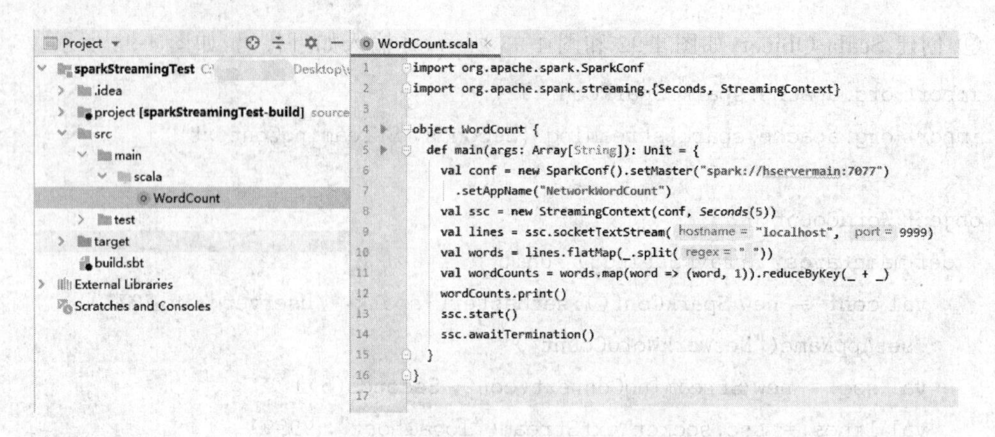

图 4-94　编写 spark-streaming 程序

⑩ 将项目打成 jar 包。选择“File→Project Structure”或按快捷键“Ctrl＋Alt＋Shift＋S”打开项目设置，并按图 4-95 和图 4-96 所示步骤执行。

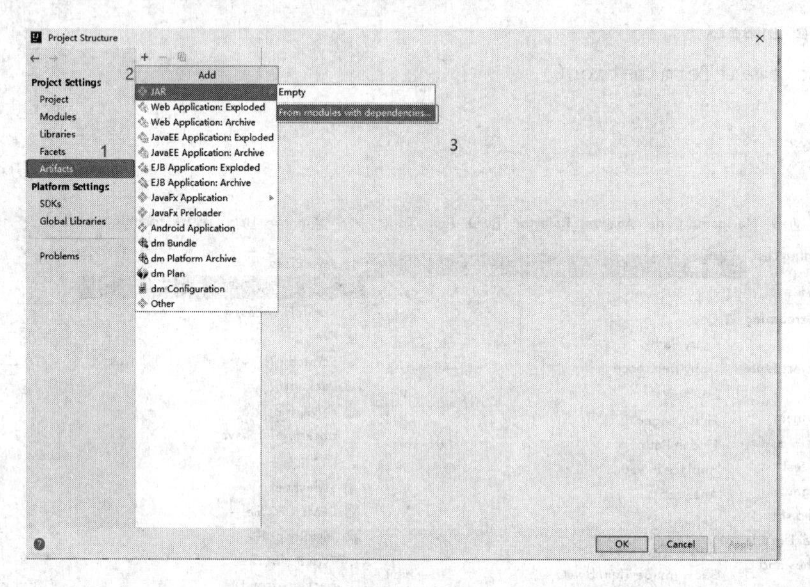

图 4-95　程序打包(一)

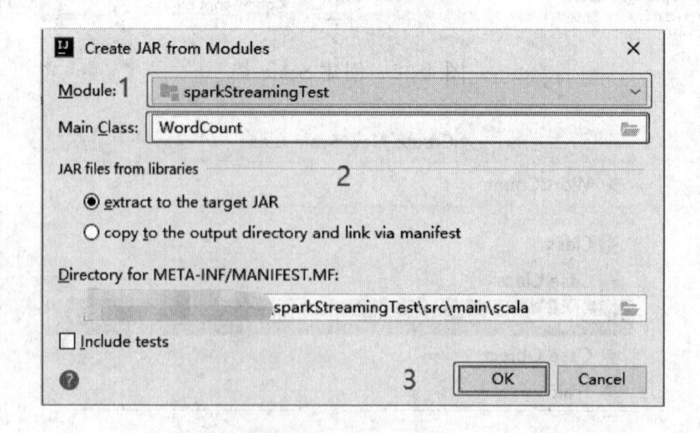

图 4-96　程序打包(二)

将所有依赖 jar 包以及 Module Library 都从打包目录中删除，只保留编译文件，如图 4-97 所示。打包目录如图 4-98 所示。

图 4-97　程序打包(三)

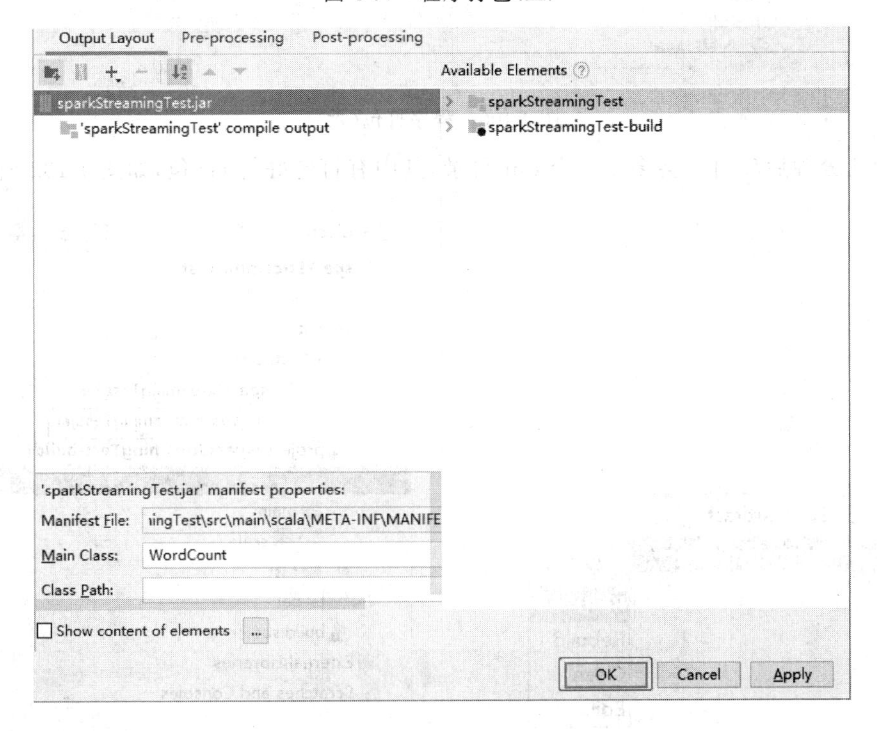

图 4-98　程序打包(四)

设置好之后会出现打包文件 MANIFEST. MF，如图 4-99 所示。

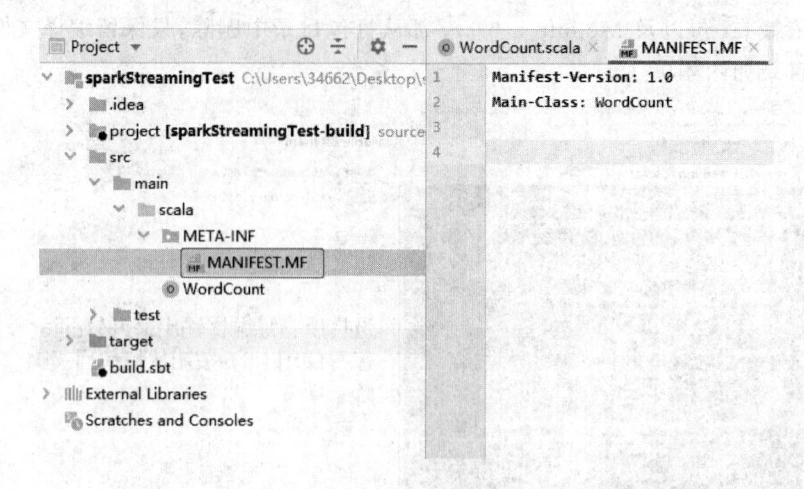

图 4-99　程序打包(五)

打包文件:选择“Build→Build Artifacts”,如图 4-100 和图 4-101 所示。

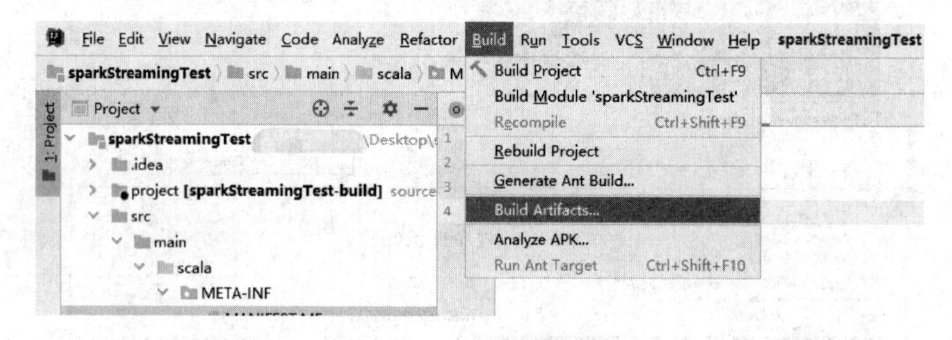

图 4-100　程序打包(六)

等待几分钟后项目中会多出一个 out 目录,其中有打包好的 jar 包,如图 4-102 所示。

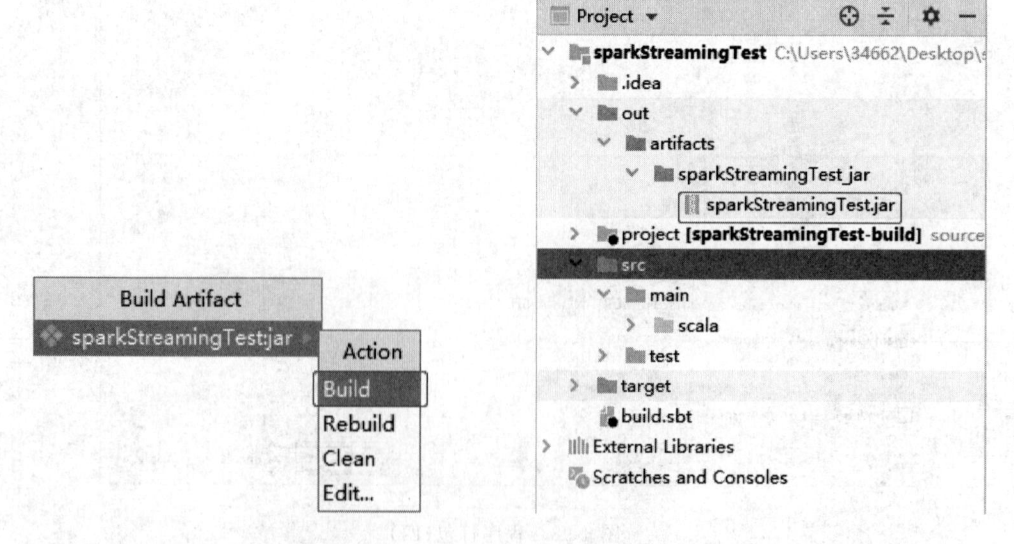

图 4-101　程序打包(七)　　　　图 4-102　程序 jar 包

⑪ 将 jar 包通过 Xftp 上传到服务器端,如图 4-103 所示。

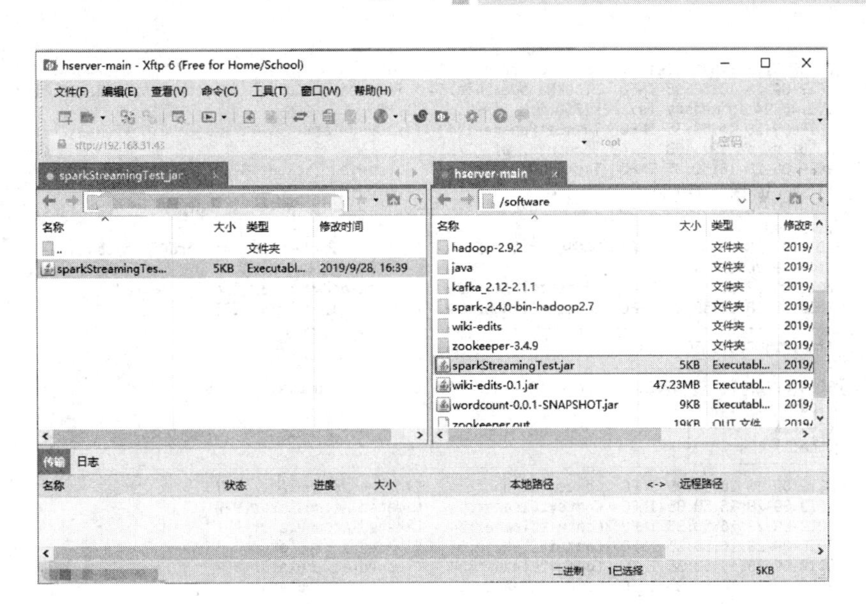

图 4-103 上传 jar 包

⑫ 在 1 号终端输入图 4-104 所示的命令。

```
[root@hservermain spark-2.4.0-bin-hadoop2.7]# ./bin/spark-submit --class WordCount ../spa
rkStreamingTest.jar
```

图 4-104 运行程序

刚才输入的消息被实时统计出了单词出现频率，再输入一些消息，观察终端状态，如图 4-105、图 4-106 和图 4-107 所示。

```
43:46494 in memory (size: 1750.0 B, free: 413.9 MB)
2019-09-28 16:49:01 INFO  BlockManagerInfo:54 - Added broadcast_5_piece0 in memory on 192
.168.31.43:46494 (size: 1750.0 B, free: 413.9 MB)
2019-09-28 16:49:01 INFO  TaskSetManager:54 - Finished task 0.0 in stage 6.0 (TID 73) in
194 ms on 192.168.31.43 (executor 0) (1/1)
2019-09-28 16:49:01 INFO  TaskSchedulerImpl:54 - Removed TaskSet 6.0, whose tasks have al
l completed, from pool
2019-09-28 16:49:01 INFO  DAGScheduler:54 - ResultStage 6 (print at WordCount.scala:12) f
inished in 0.267 s
2019-09-28 16:49:01 INFO  DAGScheduler:54 - Job 3 finished: print at WordCount.scala:12,
took 0.295909 s
-------------------------------------------
Time: 1569660540000 ms
-------------------------------------------
(hello,1)
(shell,1)
(test,1)
(world,1)
(the,1)

2019-09-28 16:49:01 INFO  JobScheduler:54 - Finished job streaming job 1569660540000 ms.0
 from job set of time 1569660540000 ms
2019-09-28 16:49:01 INFO  JobScheduler:54 - Total delay: 1.160 s for time 1569660540000 m
s (execution: 0.969 s)
2019-09-28 16:49:01 INFO  ContextCleaner:54 - Cleaned accumulator 78
2019-09-28 16:49:01 INFO  ContextCleaner:54 - Cleaned accumulator 101
2019-09-28 16:49:01 INFO  ContextCleaner:54 - Cleaned accumulator 75
```

图 4-105 程序运行结果(一)

```
[root@hservermain ~]# nc -lk 9999
hello world
test the shell
test the spark-submit
```

图 4-106 向端口写入

```
2.168.31.43:46494 (size: 1752.0 B, free: 413.9 MB)
2019-09-28 16:50:35 INFO  BlockManagerInfo:54 - Removed broadcast_40_piece0 on 192.168.31
.43:46494 in memory (size: 1750.0 B, free: 413.9 MB)
2019-09-28 16:50:35 INFO  TaskSetManager:54 - Finished task 0.0 in stage 82.0 (TID 112) i
n 161 ms on 192.168.31.43 (executor 0) (1/1)
2019-09-28 16:50:35 INFO  TaskSchedulerImpl:54 - Removed TaskSet 82.0, whose tasks have a
ll completed, from pool
2019-09-28 16:50:35 INFO  DAGScheduler:54 - ResultStage 82 (print at WordCount.scala:12)
finished in 0.256 s
2019-09-28 16:50:35 INFO  DAGScheduler:54 - Job 41 finished: print at WordCount.scala:12,
 took 0.265058 s
2019-09-28 16:50:35 INFO  ContextCleaner:54 - Cleaned accumulator 984
2019-09-28 16:50:35 INFO  ContextCleaner:54 - Cleaned accumulator 1021
-------------------------------------------
Time: 1569660635000 ms
-------------------------------------------
2019-09-28 16:50:35 INFO  ContextCleaner:54 - Cleaned accumulator 1038
(spark-submit,1)
(test,1)
(the,1)

2019-09-28 16:50:35 INFO  ContextCleaner:54 - Cleaned accumulator 1045
2019-09-28 16:50:35 INFO  ContextCleaner:54 - Cleaned accumulator 979
2019-09-28 16:50:35 INFO  ContextCleaner:54 - Cleaned accumulator 954
2019-09-28 16:50:35 INFO  ContextCleaner:54 - Cleaned accumulator 1015
2019-09-28 16:50:35 INFO  ContextCleaner:54 - Cleaned accumulator 987
2019-09-28 16:50:35 INFO  ContextCleaner:54 - Cleaned accumulator 1032
```

图 4-107　程序运行结果(二)

⑬ 在浏览器中可以查看到正在运行的任务,如图 4-108 所示。

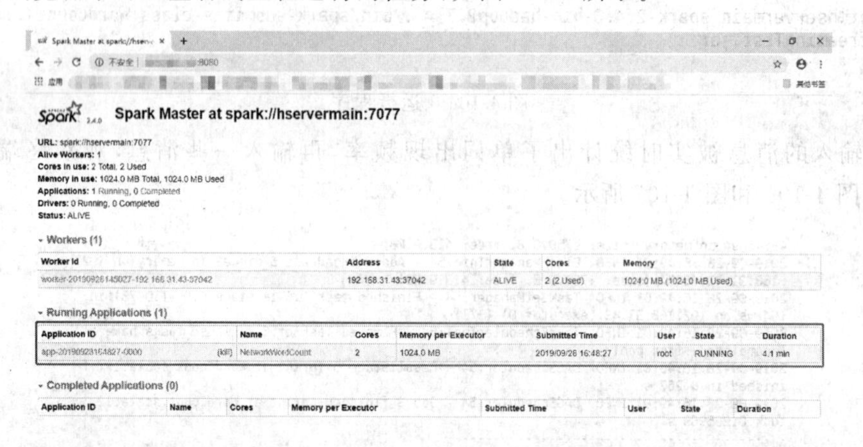

图 4-108　在管理页查看运行中的任务

7. 参考答案

实验作业的答案见本节扩展资料部分。

4.5 Flink 安装及其基本操作

1. 实验目的

学会在计算机上安装大数据计算引擎 Flink 并掌握其工作流程。

2. 实验要求

① Java 8。

② Linux、MacOSX 或其他类似于 Unix 的操作系统(不支持 Windows)。

③ 8 GB 内存。

④ 2 个 vCPU。

⑤ Eclipse 或 IDEA 等编辑器。

⑥ Kafka。

3. 预备知识

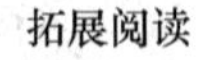

Flink 介绍

Apache Flink 是一个低延迟、高吞吐、统一的大数据计算引擎，其提供 Exactly-Once 的一致性语义保证了数据的正确性。Flink 与 Storm 和 Spark Streaming 相比具有很大优势，被誉为下一代最适用的大数据计算引擎。Flink 具备了 Spark 的所有优点，并且针对 Spark Streaming 的近似实时计算进行了改进，实现了真正意义上的实时计算，在流式计算、迭代计算与 Hadoop 兼容性等方面都要优于 Spark，其缺点在于对 SQL 的优化以及目前社区活跃度不如 Spark。

Flink 是经典的 Master-Slave 架构，在集群启动过程中，Flink 会启动一个 JobManager 进程和至少一个 TaskManager 进程。

4. 实验内容[28]

(1) 实验 1：Flink 安装与示例程序运行

① 下载 flink-1.7.2-bin-scala_2.12.tgz 安装包。

② 终端运行 tar 命令解压安装包。

```
tar -zxvf /install/flink-1.7.2-bin-scala_2.12.tgz
```

③ 启动本地 Flink 集群并在浏览器管理页上验证。

```
./bin/start-cluster.sh
```

查看 Java 进程启动情况。

```
jps
```

在浏览器中输入下述链接验证安装成功(Flink 网页端口 8081)。

```
192.168.159.130:8081
```

④ 开启新的终端窗口(2 号)，并对 9000 端口进行实时输入。

```
nc -l 9000
```

⑤ 在 1 号终端输入下述命令运行示例程序(监听 9000 端口)。

```
./bin/flink run examples/streaming/SocketWindowWordCount.jar --port 9000
```

⑥ 在浏览器管理页查看任务运行情况。

(2) 实验 2：本地开发 Flink 程序并发布到集群运行

① 在服务器端创建 Maven 程序。

```
mvn archetype:generate \
    -DarchetypeGroupId = org.apache.flink \
    -DarchetypeArtifactId = flink-quickstart-java \
    -DarchetypeVersion = 1.7.2 \
    -DgroupId = wiki-edits \
    -DartifactId = wiki-edits \
    -Dversion = 0.1 \
    -Dpackage = wikiedits \
    -DinteractiveMode = false
```

② 将程序通过 Xftp 从服务器端传到本地。

③ 将 Maven 项目导入 IDEA，删除 wikiedits 目录下的两个文件，并创建名为 WikipediaAnalysis 的文件，修改 pom.xml 内的依赖。pom.xml 内修改的依赖如下所示。

```
<dependencies>
    <dependency>
        <groupId>org.slf4j</groupId>
        <artifactId>slf4j-log4j12</artifactId>
        <version>1.7.7</version>
        <scope>runtime</scope>
    </dependency>
    <dependency>
        <groupId>org.apache.flink</groupId>
        <artifactId>flink-java</artifactId>
        <version>${flink.version}</version>
    </dependency>
    <dependency>
        <groupId>org.apache.flink</groupId>
        <artifactId>flink-streaming-java_2.11</artifactId>
        <version>${flink.version}</version>
    </dependency>
    <dependency>
        <groupId>org.apache.flink</groupId>
        <artifactId>flink-clients_2.11</artifactId>
        <version>${flink.version}</version>
    </dependency>
    <dependency>
        <groupId>org.apache.flink</groupId>
        <artifactId>flink-connector-wikiedits_2.11</artifactId>
        <version>${flink.version}</version>
    </dependency>
</dependencies>
```

④ 在 IDEA 中的 WikipediaAnalysis.class 文件中编写代码，实现对维基百科的实时修改监控（代码见扩展资料），运行并查看实时输出结果。

⑤ 在 pom 文件中添加 Kafka 依赖。

```
<dependency>
    <groupId>org.apache.flink</groupId>
    <artifactId>flink-connector-kafka-0.11_2.11</artifactId>
    <version>${flink.version}</version>
</dependency>
```

修改 Flink 程序的 Sink 源，改为 Kafka。

```
result
    .map(new MapFunction<Tuple2<String,Long>, String>() {
      @Override
      public String map(Tuple2<String, Long> tuple) {
          return tuple.toString();
      }
    })
    .addSink(new FlinkKafkaProducer011<>("localhost:9092", "wiki-result", new
            SimpleStringSchema()));
```

打包程序并上传到服务器，打包命令如下所示。

```
mvn clean package
```

⑥ 在服务器终端依次启动 Flink、ZooKeeper 以及 Kafka。

```
bin/start-cluster.sh
bin/zookeeper-server-start.shconfig/zookeeper.properties &
bin/kafka-server-start.sh -daemon config/server.properties
```

使用 jps 命令查看启动进程。

```
# jps
3762 StandalongSessionClusterEntrypoint
4292 QuorumPeerMain
4212 TaskManagerRunner
4875 Kafka
4894 Jps
```

⑦ 运行程序并实时查看 Kafka 消费情况。

创建名为 wiki-result 的 Topic：

```
bin/kafka-topics.sh --create --zookeeper localhost:2181 --replication-factor 1
--partitions 1 --topic wiki-result
```

切换到 flink 目录下执行以下程序。

```
cd /opt/flink/flink-1.7.2
bin/flink run -c wikiedits.WikipediaAnalysis ../wiki-edits-0.1.jar
```

开启新的服务器终端(2号)，使用 Kafka 消费者实时消费 Flink 程序计算结果，并观察浏览器管理页面的任务运行情况。

```
bin/kafka-console-consumer.sh --bootstrap-server localhost:9092 --topic wiki-result
```

5. 实验作业

① 下载 Flink 安装包并在虚拟机上安装，运行入门示例程序。

② 本地开发 Flink 程序并发布到 Flink 集群运行。程序描述：实时监控维基百科编辑情况，并抓取编辑者 IP 以及编辑内容的长度，抓取数据实时写入 Kafka(即被 Kafka 消费者实时消费)。

6. 扩展资料

(1) 实验1：Flink 安装与示例程序运行

① 下载 flink-1.7.2-bin-scala_2.12.tgz 安装包并解压，如图 4-109 所示。

```
[root@hservermain software]# tar -zxvf /install/flink-1.7.2-bin-scala_2.12.tgz
```

图 4-109　解压安装文件

解压后会出现蓝色的文件,如图 4-110 所示。

```
[root@hservermain software]# ls
apache-flume-1.9.0-bin  hadoop-2.9.2
apache-storm-1.2.2      java
flink-1.7.2             kafka_2.12-2.1.1
[root@hservermain software]#
```

图 4-110　查看安装结果

② 启动本地 Flink 集群并在网页端验证,如图 4-111 所示。

```
[root@hservermain flink-1.7.2]# ./bin/start-cluster.sh
```

图 4-111　启动 Flink

使用 jps 命令查看进程启动情况,如图 4-112 所示。

```
[root@hservermain flink-1.7.2]# jps
29555 StandaloneSessionClusterEntrypoint
29989 TaskManagerRunner
30008 Jps
[root@hservermain flink-1.7.2]#
```

图 4-112　进程列表

③ 在浏览器中输入"主机 IP:端口(8081)"查看集群启动情况,如图 4-113 所示。

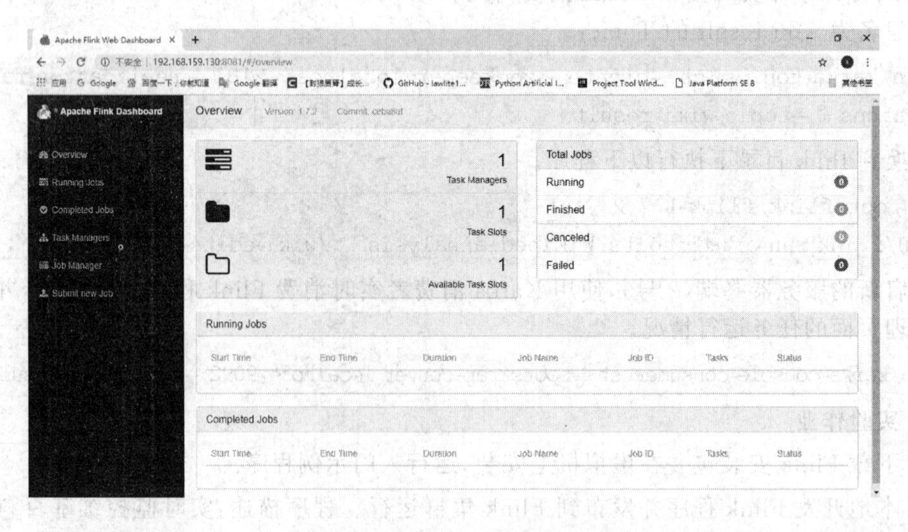

图 4-113　Flink 管理页

④ 打开新的终端窗口(2 号),对 9000 端口进行实时写入操作,如图 4-114 所示。

```
[root@hservermain software]# nc -l 9000
```

图 4-114　开启端口

同时，在 1 号终端启动示例程序（对 9000 端口实时监控），如图 4-115 所示。

```
[root@hservermain flink-1.7.2]# ./bin/flink run examples/streaming/SocketWindowWordCount
.jar --port 9000
Starting execution of program
```

图 4-115　启动示例程序

启动任务后，在浏览器管理页面可以查看到正在运行的任务，如图 4-116 所示。

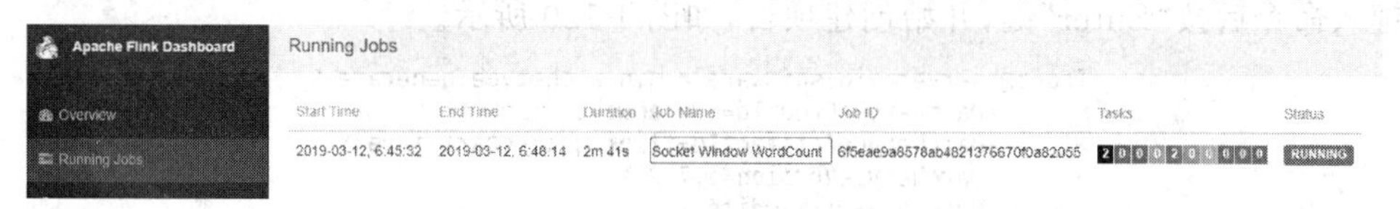

图 4-116　在管理页查看运行中的任务

在 2 号终端向 9000 端口输入消息，如图 4-117 所示，并在网页端查看程序的实时运行状况。

```
[root@hservermain software]# nc -l 9000
hello world hello flink
```

图 4-117　向端口写入

在 Task Managers 处可以查看到任务的管理情况，如图 4-118 所示。

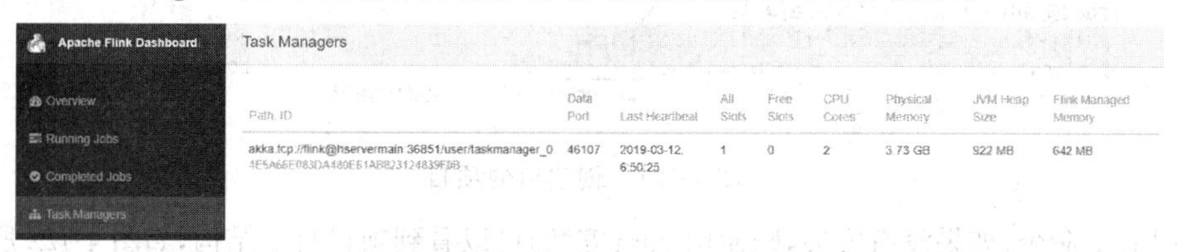

图 4-118　Task Managers 处查看任务

在 Metrics 处可以看到资源的使用情况，Logs 处可以看到任务运行过程中产生的日志信息，Stdout 处可以查看程序实时运行的结果，如图 4-119 所示。

图 4-119　任务运行日志

（2）实验 2：本地开发 Flink 程序并发布到集群运行

① 使用 Flink Maven 原型在服务器端创建 Maven 项目，切换到/software 目录下，输入下述命令。

```
mvn archetype:generate \
    -DarchetypeGroupId = org.apache.flink \
    -DarchetypeArtifactId = flink-quickstart-java \
    -DarchetypeVersion = 1.7.2 \
```

```
-DgroupId = wiki-edits \
-DartifactId = wiki-edits \
-Dversion = 0.1 \
-Dpackage = wikiedits \
-DinteractiveMode = false
```

键入命令后按“Enter”键，开始创建项目，如图 4-120 所示。

```
[root@hservermain software]# mvn archetype:generate \
>     -DarchetypeGroupId=org.apache.flink \
>     -DarchetypeArtifactId=flink-quickstart-java \
>     -DarchetypeVersion=1.7.2 \
>     -DgroupId=wiki-edits \
>     -DartifactId=wiki-edits \
>     -Dversion=0.1 \
>     -Dpackage=wikiedits \
>     -DinteractiveMode=false
```

图 4-120　创建 Maven 项目

项目创建完成后在/software 目录下会多出一个名为 wiki-edits 的 Maven 项目，如图 4-121 所示。

```
[root@hservermain software]# ls
apache-flume-1.9.0-bin  flink-1.7.2   kafka_2.12-2.1.1                  zookeeper-3.4.9
apache-maven-3.6.2      hadoop-2.9.2  wiki-edits                        zookeeper.out
apache-storm-1.2.2      java          wordcount-0.0.1-SNAPSHOT.jar
```

图 4-121　创建后的项目

使用 tree 命令(如果没有安装，则使用 yum 安装)可以看到项目目录结构，如图 4-122 所示。

```
[root@hservermain software]# tree wiki-edits/
wiki-edits/
├── pom.xml
└── src
    └── main
        ├── java
        │   └── wikiedits
        │       ├── BatchJob.java
        │       └── StreamingJob.java
        └── resources
            └── log4j.properties

5 directories, 4 files
[root@hservermain software]#
```

图 4-122　项目目录结构

② 通过 Xftp 将项目从服务器导入本地，如图 4-123 所示。

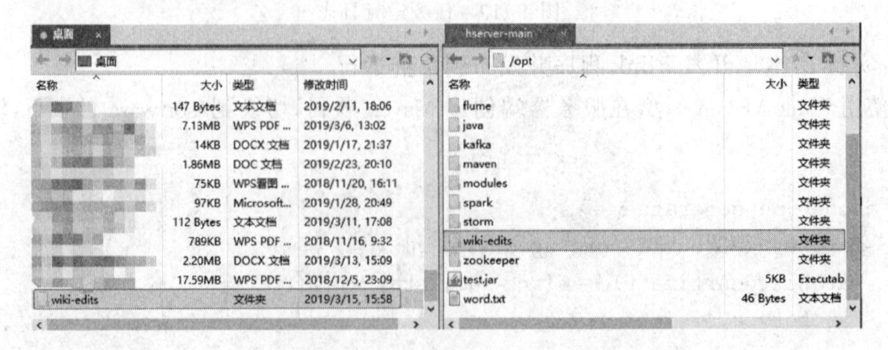

图 4-123　将程序导入本地

打开 IDEA，通过“File→Open”打开项目，如图 4-124 所示。

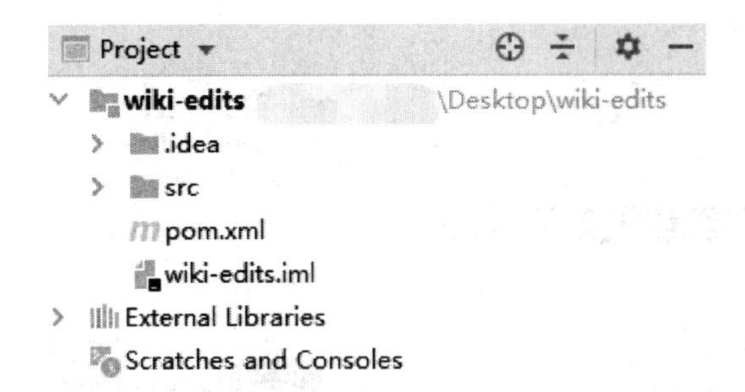

图 4-124 IDEA 打开项目

修改 pom. xml 的依赖(将 pom 文件底层的依赖删除，合并到一起)，如下所示。

```
<dependencies>
    <dependency>
      <groupId>org.slf4j</groupId>
      <artifactId>slf4j-log4j12</artifactId>
      <version>1.7.7</version>
      <scope>runtime</scope>
    </dependency>
    <dependency>
        <groupId>org.apache.flink</groupId>
        <artifactId>flink-java</artifactId>
        <version>${flink.version}</version>
    </dependency>
    <dependency>
        <groupId>org.apache.flink</groupId>
        <artifactId>flink-streaming-java_2.11</artifactId>
        <version>${flink.version}</version>
    </dependency>
    <dependency>
        <groupId>org.apache.flink</groupId>
        <artifactId>flink-clients_2.11</artifactId>
        <version>${flink.version}</version>
    </dependency>
    <dependency>
        <groupId>org.apache.flink</groupId>
        <artifactId>flink-connector-wikiedits_2.11</artifactId>
        <version>${flink.version}</version>
    </dependency>
</dependencies>
```

删除 wikiedits 目录下的两个文件，并创建名为 WikipediaAnalysis 的 Java 文件，如图 4-125 和图 4-126 所示。

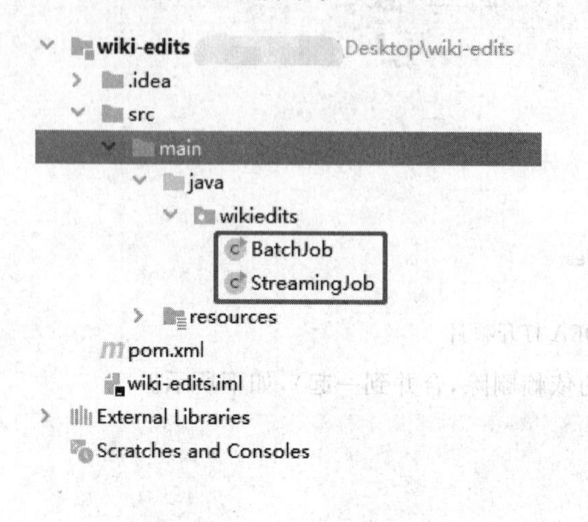

图 4-125 删除项目下文件

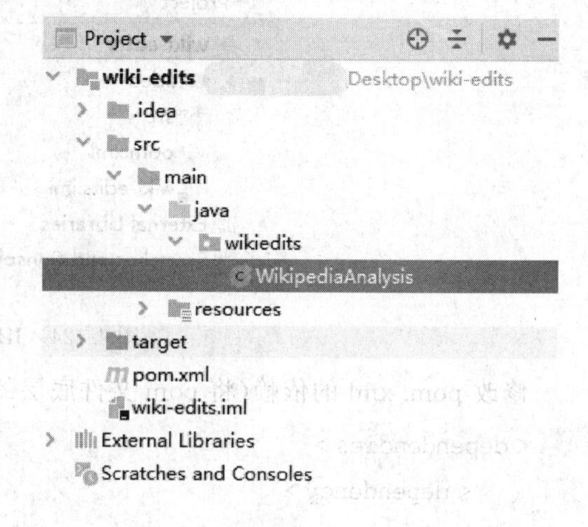

图 4-126 在项目下创建新的文件

添加代码到 WikipediaAnalysis 文件中。

```
package wikiedits;

import org.apache.flink.api.common.functions.FoldFunction;
import org.apache.flink.api.java.functions.KeySelector;
import org.apache.flink.api.java.tuple.Tuple2;
import org.apache.flink.streaming.api.datastream.DataStream;
import org.apache.flink.streaming.api.datastream.KeyedStream;
import org.apache.flink.streaming.api.environment.StreamExecutionEnvironment;
import org.apache.flink.streaming.api.windowing.time.Time;
import org.apache.flink.streaming.connectors.wikiedits.WikipediaEditEvent;
import org.apache.flink.streaming.connectors.wikiedits.WikipediaEditsSource;
public class WikipediaAnalysis {
  public static void main(String[] args) throws Exception {
     StreamExecutionEnvironment see = StreamExecutionEnvironment.
                                      getExecutionEnvironment();
     DataStream<WikipediaEditEvent> edits = see.addSource(new
                                            WikipediaEditsSource());
     KeyedStream<WikipediaEditEvent, String> keyedEdits = edits
          .keyBy(new KeySelector<WikipediaEditEvent, String>() {
             @Override
             public String getKey(WikipediaEditEvent event) {
                return event.getUser();
             }
          });
```

```
        DataStream<Tuple2<String, Long>> result = keyedEdits
                .timeWindow(Time.seconds(5))
                .fold(new Tuple2<>("", 0L), new FoldFunction<WikipediaEditEvent,
                        Tuple2<String, Long>>() {
                    @Override
                    public Tuple2<String, Long> fold(Tuple2<String, Long> acc,
                                                     WikipediaEditEvent event) {
                        acc.f0 = event.getUser();
                        acc.f1 += event.getByteDiff();
                        return acc;
                    }
                });
        result.print();
        see.execute();
    }
}
```

在 WikipediaAnalysis 文件中运行上述程序，待初始化日志输出后，程序会输出图 4-127 所示的信息。

```
10:43:23,769 INFO  org.apache.flink.runtime.state.heap.HeapKeyedStateBackend      - Initializing heap keyed state backend with stream factory.
10:43:23,769 INFO  org.apache.flink.runtime.state.heap.HeapKeyedStateBackend      - Initializing heap keyed state backend with stream factory.
10:43:23,772 INFO  org.apache.flink.runtime.state.heap.HeapKeyedStateBackend      - Initializing heap keyed state backend with stream factory.
10:43:23,772 INFO  org.apache.flink.runtime.state.heap.HeapKeyedStateBackend      - Initializing heap keyed state backend with stream factory.
2> (BrownHairedGirl,-27)
3> (WP 1.0 bot,0)
3> (AnomieBOT,20)
4> (00:1700:CD00:5060:F4F8:FC75:AF44:9714,21)
4> (01:7400:6004:35BE:807F:92EE:DE6C:2C2C,-27)
4> (Ncboss,-3)
4> (RickyLol,-91)
4> (Liz,149)
4> (00:1700:8B60:CA00:612D:B277:436A:A222,-10)
4> (Monkbot,17)
1> (Khoshhat,368)
1> (Egsan Bacon,-6)
1> (Salavat,231)
1> (KConWiki,26)
```

图 4-127 程序运行结果

该程序实时统计修改维基百科的用户的 IP 以及修改字数。需要注意的是，当程序出现连接失败的字样时，要检查一下网络是否可以连接到维基百科。

③ 在 pom 中添加 Flink 的 Kafka 接口依赖。

```
<dependency>
    <groupId>org.apache.flink</groupId>
    <artifactId>flink-connector-kafka-0.11_2.11</artifactId>
    <version>${flink.version}</version>
</dependency>
```

修改 WikipediaAnalysis 文件，相关类需要导入，如图 4-128 所示。

```
WikipediaAnalysis.java
1  package wikiedits;
2
3  import org.apache.flink.streaming.connectors.kafka.FlinkKafkaProducer011;
4  import org.apache.flink.api.common.serialization.SimpleStringSchema;
5  import org.apache.flink.api.common.functions.MapFunction;
6  import org.apache.flink.api.common.functions.FoldFunction;
7  import org.apache.flink.api.java.functions.KeySelector;
8  import org.apache.flink.api.java.tuple.Tuple2;
9  import org.apache.flink.streaming.api.datastream.DataStream;
10 import org.apache.flink.streaming.api.datastream.KeyedStream;
11 import org.apache.flink.streaming.api.environment.StreamExecutionEnvironment;
12 import org.apache.flink.streaming.api.windowing.time.Time;
13 import org.apache.flink.streaming.connectors.wikiedits.WikipediaEditEvent;
14 import org.apache.flink.streaming.connectors.wikiedits.WikipediaEditsSource;
15
16 public class WikipediaAnalysis {
```

图 4-128　在程序中导入包

把 Sink 从 print 修改成 Kafka，如图 4-129 所示。

```
//      result.print();
    result
            .map(new MapFunction<Tuple2<String,Long>, String>() {
                @Override
                public String map(Tuple2<String, Long> tuple) {
                    return tuple.toString();
                }
            })
            .addSink(new FlinkKafkaProducer011<>( brokerList: "localhost:9092",  topicId: "wiki-result", new SimpleStringSchema()));
    see.execute();
    }
}
```

图 4-129　修改 Sink

使用 Maven 将项目打包，如图 4-130 所示。

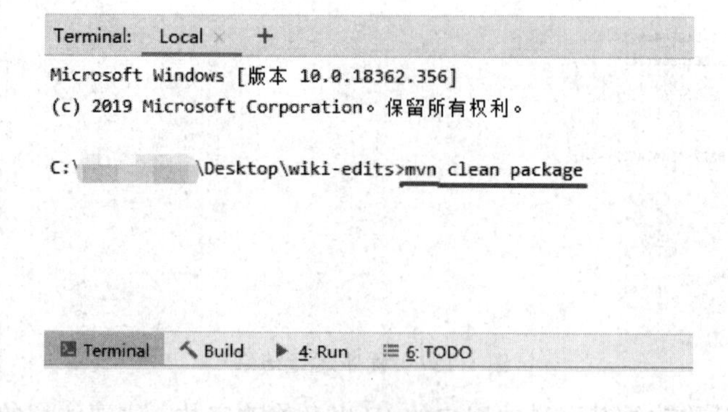

图 4-130　项目打包

等待几分钟后在 target 目录下会生成 wiki-edits-0.1.jar，如图 4-131 所示。

通过 Xftp 将文件传到服务器端，如图 4-132 所示。

④ 在服务器终端依次启动 Flink、ZooKeeper 以及 Kafka。

在 flink 目录下输入以下命令，如图 4-133 所示。

```
bin/start-cluster.sh
```

切换到 kafka 目录下，输入以下命令使 ZooKeeper 在后台运行，如图 4-134 所示。

```
bin/zookeeper-server-start.sh config/zookeeper.properties &
```

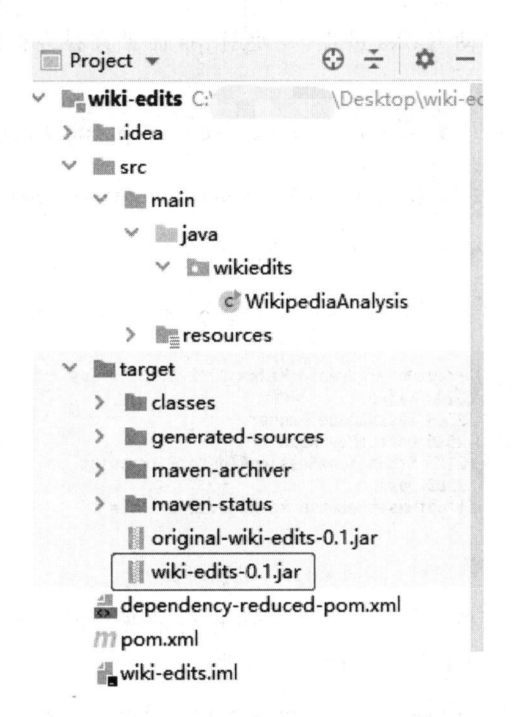

图 4-131　程序 jar 包

hserver-main

/software

名称	大小	类型	修改时
flink-1.7.2		文件夹	2019/
hadoop-2.9.2		文件夹	2019/
java		文件夹	2019/
kafka_2.12-2.1.1		文件夹	2019/
wiki-edits		文件夹	2019/
zookeeper-3.4.9		文件夹	2019/
wiki-edits-0.1.jar	47.23MB	Executabl...	2019/
wordcount-0.0.1-SNAPSHOT.jar	9KB	Executabl...	2019/
zookeeper.out	19KB	OUT 文件	2019/

图 4-132　将程序上传到服务器

```
[root@hservermain flink-1.7.2]# bin/start-cluster.sh
Starting cluster.
Starting standalonesession daemon on host hservermain.
Starting taskexecutor daemon on host hservermain.
```

图 4-133　启动 Flink

```
[root@hservermain kafka_2.12-2.1.1]# bin/zookeeper-server-start.sh config/zookeeper.prop
erties &
[1] 2636
[root@hservermain kafka_2.12-2.1.1]#
```

图 4-134　启动 ZooKeeper

输入以下命令在后台启动 Kafka 进程，并使用 jps 命令查看进程启动情况，如图 4-135 和图 4-136 所示。

```
bin/kafka-server-start.sh -daemon config/server.properties
```

```
[root@hservermain kafka_2.12-2.1.1]# bin/kafka-server-start.sh -daemon config/server.pro
perties
[root@hservermain kafka_2.12-2.1.1]#
```

图 4-135 启动 Kafka

```
[root@hservermain kafka_2.12-2.1.1]# jps
3200 Kafka
2563 TaskManagerRunner
2636 QuorumPeerMain
2127 StandaloneSessionClusterEntrypoint
3263 Jps
[root@hservermain kafka_2.12-2.1.1]#
```

图 4-136 进程列表

⑤ 在 Flink 集群运行程序并实时传到 Kafka 进行消费。创建名为 wiki-result（与程序设定需要相同）的 Topic，输入以下命令，如图 4-137 所示。

```
bin/kafka-topics.sh --create --zookeeper localhost:2181 --replication-factor 1 --partitions 1 --topic wiki-result
```

```
[root@hservermain kafka_2.12-2.1.1]# bin/kafka-topics.sh --create --zookeeper localhost:
2181 --replication-factor 1 --partitions 1 --topic wiki-result
[2019-09-28 11:08:20,943] INFO Accepted socket connection from /192.168.31.43:59048 (org
.apache.zookeeper.server.NIOServerCnxnFactory)
[2019-09-28 11:08:20,988] INFO Client attempting to establish new session at /192.168.31
.43:59048 (org.apache.zookeeper.server.ZooKeeperServer)
[2019-09-28 11:08:21,036] INFO Established session 0x1000051b2ab0001 with negotiated tim
eout 30000 for client /192.168.31.43:59048 (org.apache.zookeeper.server.ZooKeeperServer)
[2019-09-28 11:08:23,355] INFO Got user-level KeeperException when processing sessionid:
0x1000051b2ab0001 type:setData cxid:0x4 zxid:0xa5 txntype:-1 reqpath:n/a Error Path:/con
fig/topics/wiki-result Error:KeeperErrorCode = NoNode for /config/topics/wiki-result (or
g.apache.zookeeper.server.PrepRequestProcessor)
Created topic "wiki-result".
[2019-09-28 11:08:23,785] INFO Processed session termination for sessionid: 0x1000051b2a
b0001 (org.apache.zookeeper.server.PrepRequestProcessor)
[2019-09-28 11:08:23,812] INFO Closed socket connection for client /192.168.31.43:59048
which had sessionid 0x1000051b2ab0001 (org.apache.zookeeper.server.NIOServerCnxn)
[root@hservermain kafka_2.12-2.1.1]#
```

图 4-137 创建 Topic

切换到 flink 目录下，在 wiki-result 的 Topic（程序中已经设置）上运行程序，输入以下命令，如图 4-138 所示。

```
bin/flink run -c wikiedits.WikipediaAnalysis ../wiki-edits-0.1.jar
```

```
[root@hservermain flink-1.7.2]# bin/flink run -c wikiedits.WikipediaAnalysis ../wiki-edi
ts-0.1.jar
Starting execution of program
```

图 4-138 运行 Flink 程序

⑥ 在 Flink 浏览器管理页面查看程序运行情况以及 Kafka 消费情况。在浏览器中输入"主机 IP:Flink 端口(8081)"，查看到有任务正在运行，如图 4-139 所示。

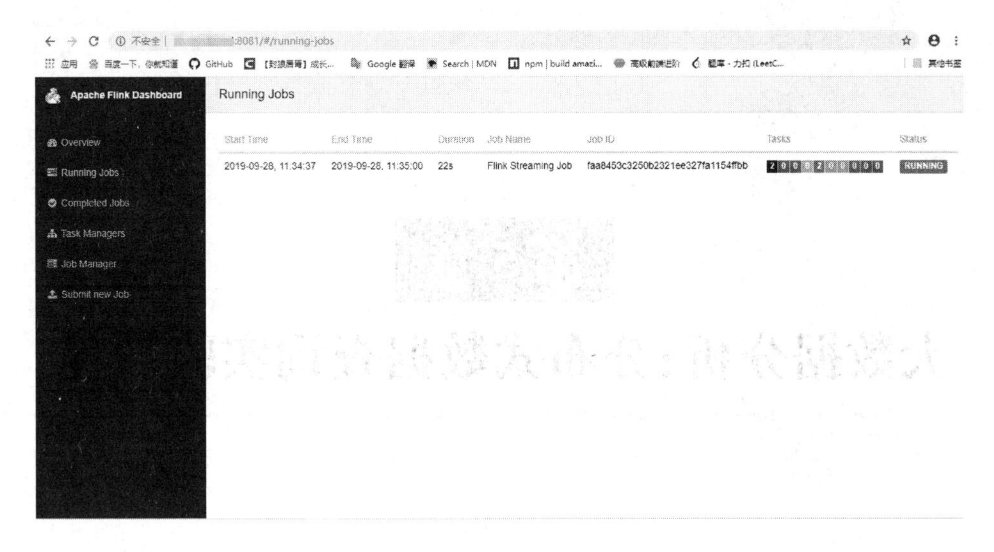

图 4-139 在管理页查看程序运行情况

打开一个新的终端窗口(2 号)，切换到 kafka 目录下，输入以下命令，实时消费 Flink 计算结果数据，如图 4-140 所示。

```
bin/kafka-console-consumer. sh --bootstrap-server localhost: 9092 --topic wiki-result
```

```
[root@hservermain kafka_2.12-2.1.1]# bin/kafka-console-consumer.sh --bootstrap-server loc
alhost:9092 --topic wiki-result
(WP 1.0 bot,4133)
(Vmavanti,-25)
(Jevansen,147)
(Dl2000,30)
(1.178.234.183,-19)
(Seligne,244)
(05:9800:BC11:BD0D:DDA7:8F2A:C48:CA52,398)
(Monkbot,5)
(Vallee01,88)
(Kin no Ryū,2)
```

图 4-140 使用命令查看程序运行情况

7. 参考答案

实验作业的答案见本节扩展资料部分。

第5章

大数据分析:分布式数据查询实验教程

5.1 Hive的数据导入与数据查询

1. 实验目的

① 学会使用相关命令将数据导入 Hive 分布式数据仓库中。

② 学会使用 SQL 语句对 Hive 分布式数据仓库中的数据进行查询。

2. 实验要求

① Java 8。

② 搭建好 Hadoop 集群的计算机。

③ 安装了 Hive 分布式数据仓库的计算机。

④ MySQL 数据库。

3. 预备知识

Hive 是一个基于 Hadoop 的数据仓库工具,可以将结构化的数据文件映射为一张数据库表,并提供简单的 SQL 查询功能,可以将 SQL 语句转换为 MapReduce 任务运行。Hive 的优点是学习成本低,可以通过类 SQL 语句快速实现简单的 MapReduce 统计,不必开发专门的 MapReduce 应用,十分适用于数据仓库的统计分析。Hive 并不适合那些需要低延迟的应用,如联机事务处理(OLTP)。Hive 查询操作过程严格遵守 Hadoop MapReduce 的作业执行模型,整个查询过程也比较慢,不适合实时数据分析。Hive 的最佳使用场合是大数据集的批处理作业,如网络日志分析。

4. 实验内容

(1) 数据导入[29]

① 将本地文件系统中的数据导入 Hive 中。首先进入 Hive 数据仓库中,在 Hive 中建好与导入的数据格式一致的 test 表,建表的命令如下所示。

```
create table test
(id int,
name string,
subject string,
```

```
 grade int
)
ROW FORMAT DELIMITED
FIELDS TERMINATED BY '\t'
STORED AS TEXTFILE;
```

在本地文件系统的 home 目录下建立 test. txt 文件，内容如下所示。

```
1    Tony     Chinese    89
2    Alice    Science    87
3    Mark     Physics    85
```

test. txt 文件中的数据列之间是使用“\t”分割的，可以使用以下命令将这个文件中的数据导入 test 表中。

```
load data local inpath "/home/test.txt" into table test;
```

② 将分布式文件系统 HDFS 上的数据导入 Hive 中。将数据从本地文件系统导入 Hive 中，实质是先将数据复制到 HDFS 的临时目录下，再将数据从临时目录下移动到对应的 Hive 表的目录下。因此，Hive 支持将数据直接从 HDFS 的某个目录下移动到相应 Hive 表的目录下。将本地的 home 目录下的文件 test. txt 传到 HDFS 的 home 目录下，具体的操作如下所示。

```
hdfs dfs -put /home/test.txt /home
hdfs dfs -cat /home/test.txt
1    Tony     Chinese    89
2    Alice    Science    87
3    Mark     Physics    85
```

以上是需要导入 Hive 中的数据，这个文件存放在 HDFS 的 home 目录下。与①相同，需要先在 Hive 中建立与导入数据格式对应的 test 表，然后使用以下命令将 test. txt 中的数据导入建好的 test 表中。

```
load data inpath '/home/test.txt' into table test;
```

(2) 数据查询

以上述导入 Hive 数据仓库中的 test 表为例，进行查询操作。

① 查询 test 表的所有数据，操作如下所示。

```
select * from test;
test.id  test.name  test.subject  test.grade
   1        Tony       Chinese         89
   2        Alice      Science         87
   3        Mark       Physics         85
```

拓展阅读

HiveQL 查询

② 增加 test 表的数据，操作如下所示。

```
insert into test values (4,'Grace','Math',86);
select * from test;
test.id   test.name   test.subject   test.grade
   4        Grace        Math            86
   1        Tony        Chinese          89
   2        Alice       Science          87
   3        Mark        Physics          85
```

③ 删除 test 表的数据，操作如下所示。

```
insert overwrite table test select * from test where id ! =2;
select * from test;
test.id   test.name   test.subject   test.grade
   1        Tony        Chinese          89
   3        Mark        Physics          85
   4        Grace        Math            86
```

④ 使用 where 语句对 test 表进行查询，操作如下所示。

```
select * from test where name='Mark';
test.id   test.name   test.subject   test.grade
   3        Mark        Physics          85
```

⑤ 排序。根据 grade 列对 test 表的数据进行降序排列，操作如下所示。

```
select * from test order by grade desc;
test.id   test.name   test.subject   test.grade
   1        Tony        Chinese          89
   4        Grace        Math            86
   3        Mark        Physics          85
```

5. 实验作业

① 了解 Hive 分布式数据仓库的架构和工作原理。

② 完成从本地文件系统中导入数据到 Hive 表和从 HDFS 上导入数据到 Hive 表。

③ 对导入 Hive 分布式数据仓库的数据进行查询。

6. 扩展资料

(1) 数据导入

① 将本地文件系统中的数据导入 Hive 中。首先进入 Hive 数据仓库中，具体操作如图 5-1 所示。

```
[root@slave2 home]# hive
SLF4J: Class path contains multiple SLF4J bindings.
SLF4J: Found binding in [jar:file:/usr/hdp/3.0.1.0-187/hive/lib/log4j-slf4j-impl
-2.10.0.jar!/org/slf4j/impl/StaticLoggerBinder.class]
SLF4J: Found binding in [jar:file:/usr/hdp/3.0.1.0-187/hadoop/lib/slf4j-log4j12-
1.7.25.jar!/org/slf4j/impl/StaticLoggerBinder.class]
SLF4J: See http://www.slf4j.org/codes.html#multiple_bindings for an explanation.
SLF4J: Actual binding is of type [org.apache.logging.slf4j.Log4jLoggerFactory]
Connecting to jdbc:hive2://slave2.hadoop:2181,slave4.hadoop:2181,slave5.hadoop:2
181/default;password=password;serviceDiscoveryMode=zooKeeper;user=hdfs;zooKeeper
Namespace=hiveserver2
19/09/25 11:26:59 [main]: INFO jdbc.HiveConnection: Connected to slave2.hadoop:1
0000
Connected to: Apache Hive (version 3.1.0.3.0.1.0-187)
Driver: Hive JDBC (version 3.1.0.3.0.1.0-187)
Transaction isolation: TRANSACTION_REPEATABLE_READ
Beeline version 3.1.0.3.0.1.0-187 by Apache Hive
0: jdbc:hive2://slave2.hadoop:2181,slave4.had>
```

图 5-1 进入 Hive

在 Hive 中建好与导入数据格式一致的 test 表，命令和建表结果如图 5-2 所示。

```
0: jdbc:hive2://slave2.hadoop:2181,slave4.had> create table test
. . . . . . . . . . . . . . . . . . . . . . .> (id int,
. . . . . . . . . . . . . . . . . . . . . . .> name string,
. . . . . . . . . . . . . . . . . . . . . . .> subject string,
. . . . . . . . . . . . . . . . . . . . . . .> grade int
. . . . . . . . . . . . . . . . . . . . . . .> )
. . . . . . . . . . . . . . . . . . . . . . .> ROW FORMAT DELIMITED
. . . . . . . . . . . . . . . . . . . . . . .> FIELDS TERMINATED BY '\t'
. . . . . . . . . . . . . . . . . . . . . . .> STORED AS TEXTFILE;

INFO  : Starting task [Stage-0:DDL] in serial mode
INFO  : Completed executing command(queryId=hive_20190925133728_1a9a0a7f-f2a7-44
8e-9c03-d911861fce07); Time taken: 0.262 seconds
INFO  : OK
```

图 5-2 在 Hive 中建立 test 表

在本地文件系统的 home 目录下建立 test. txt 文件，内容如图 5-3 所示。

```
[root@slave2 home]# cat test.txt
1    Tony    Chinese    89
2    Alice    Science    87
3    Mark    Physics    85
```

图 5-3 在本地建立 test. txt

通过导入命令语句将 test. txt 中的数据导入 Hive 的 test 表中，具体操作如图 5-4 所示。

```
0: jdbc:hive2://slave2.hadoop:2181,slave4.had>  load data local inpath "/home/te
st.txt" into table test;
INFO  : Compiling command(queryId=hive_20190925134915_95fdd623-4fbf-4d25-a07a-03
984329b79b): load data local inpath "/home/test.txt" into table test
INFO  : Semantic Analysis Completed (retrial = false)
INFO  : Returning Hive schema: Schema(fieldSchemas:null, properties:null)
INFO  : Completed compiling command(queryId=hive_20190925134915_95fdd623-4fbf-4d
25-a07a-03984329b79b); Time taken: 0.109 seconds
INFO  : Executing command(queryId=hive_20190925134915_95fdd623-4fbf-4d25-a07a-03
984329b79b): load data local inpath "/home/test.txt" into table test
INFO  : Starting task [Stage-0:MOVE] in serial mode
INFO  : Loading data to table default.test from file:/home/test.txt
INFO  : Starting task [Stage-1:STATS] in serial mode
INFO  : Completed executing command(queryId=hive_20190925134915_95fdd623-4fbf-4d
25-a07a-03984329b79b); Time taken: 0.365 seconds
INFO  : OK
```

图 5-4 将 test. txt 中的数据导入 test 表中

② 将分布式文件系统 HDFS 上的数据导入 Hive 中。将本地的 home 目录下的文件 test. txt 传到 HDFS 的 home 目录下，具体的操作如图 5-5 所示。

```
[root@slave2 ~]# hdfs dfs -put /home/test.txt  /home
[root@slave2 ~]# hdfs dfs  -cat /home/test.txt;
1       Tony    Chinese 89
2       Alice   Science 87
3       Mark    Physics 85
```

图 5-5 将本地的 text. txt 导入 HDFS 中

与①相同，需要在 Hive 中建立与导入数据格式一致的 test 表，然后将 HDFS 中 test. txt 的数据导入 Hive 中的 test 表中，具体操作如图 5-6 所示。

```
0: jdbc:hive2://slave2.hadoop:2181,slave4.had> load data inpath '/home/test.txt'
 into table test;
INFO  : Compiling command(queryId=hive_20190925140621_6d6ad0e3-6d90-4614-b93a-f0
3f27d976e4): load data inpath '/home/test.txt' into table test
INFO  : Semantic Analysis Completed (retrial = false)
INFO  : Returning Hive schema: Schema(fieldSchemas:null, properties:null)
INFO  : Completed compiling command(queryId=hive_20190925140621_6d6ad0e3-6d90-46
14-b93a-f03f27d976e4); Time taken: 0.131 seconds
INFO  : Executing command(queryId=hive_20190925140621_6d6ad0e3-6d90-4614-b93a-f0
3f27d976e4): load data inpath '/home/test.txt' into table test
INFO  : Starting task [Stage-0:MOVE] in serial mode
INFO  : Loading data to table default.test from hdfs://slave2.hadoop:8020/home/t
est.txt
INFO  : Starting task [Stage-1:STATS] in serial mode
INFO  : Completed executing command(queryId=hive_20190925140621_6d6ad0e3-6d90-46
14-b93a-f03f27d976e4); Time taken: 0.62 seconds
INFO  : OK
```

图 5-6 将 HDFS 中 test. txt 的数据导入 test 表中

从上面的执行结果可以看到，数据已经导入 test 表中，注意“load data inpath‘/home/test. txt’ into table test;”中是没有“local”这个词的。

(2) 数据查询

以上述导入 Hive 数据仓库中的 test 表为例，进行查询操作。

① 查询 test 表的数据，操作如图 5-7 所示。

```
0: jdbc:hive2://slave2.hadoop:2181,slave4.had> select * from test;
INFO  : Compiling command(queryId=hive_20190925140946_3000cd9b-1e16-4b10-8393-ee
6f62a980dc): select * from test
INFO  : Semantic Analysis Completed (retrial = false)
INFO  : Returning Hive schema: Schema(fieldSchemas:[FieldSchema(name:test.id, ty
pe:int, comment:null), FieldSchema(name:test.name, type:string, comment:null), F
ieldSchema(name:test.subject, type:string, comment:null), FieldSchema(name:test.
grade, type:int, comment:null)], properties:null)
INFO  : Completed compiling command(queryId=hive_20190925140946_3000cd9b-1e16-4b
10-8393-ee6f62a980dc); Time taken: 0.299 seconds
INFO  : Executing command(queryId=hive_20190925140946_3000cd9b-1e16-4b10-8393-ee
6f62a980dc): select * from test
INFO  : Completed executing command(queryId=hive_20190925140946_3000cd9b-1e16-4b
10-8393-ee6f62a980dc); Time taken: 0.01 seconds
INFO  : OK
+----------+------------+---------------+-------------+
| test.id  | test.name  | test.subject  | test.grade  |
+----------+------------+---------------+-------------+
| 1        | Tony       | Chinese       | 89          |
| 2        | Alice      | Science       | 87          |
| 3        | Mark       | Physics       | 85          |
+----------+------------+---------------+-------------+
```

图 5-7 test 表的查询结果

② 增加 test 表的数据，进行插入操作，如图 5-8 所示。

```
0: jdbc:hive2://slave2.hadoop:2181,slave4.had> insert into test values (4,'Grac
e','Math',86);

INFO  : Starting task [Stage-2:DEPENDENCY_COLLECTION] in serial mode
INFO  : Starting task [Stage-0:MOVE] in serial mode
INFO  : Loading data to table default.test from hdfs://slave2.hadoop:8020/wareho
use/tablespace/managed/hive/test/.hive-staging_hive_2019-09-25_14-33-33_741_2778
188333230944854-1/-ext-10000
INFO  : Starting task [Stage-3:STATS] in serial mode
INFO  : Completed executing command(queryId=hive_20190925143333_3d622f3d-8f7e-4c
48-b076-66d3ea941f6f); Time taken: 14.13 seconds
INFO  : OK
No rows affected (14.708 seconds)
```

图 5-8 将数据插入 test 表

插入之后，进行查询操作，如图 5-9 所示。

test.id	test.name	test.subject	test.grade
4	Grace	Math	86
1	Tony	Chinese	89
2	Alice	Science	87
3	Mark	Physics	85

图 5-9 插入操作后的查询结果

③ 删除 test 表的数据，操作如图 5-10 所示。

```
0: jdbc:hive2://slave2.hadoop:2181,slave4.had> insert overwrite table test selec
t * from test where id !=2;

INFO  : Starting task [Stage-2:DEPENDENCY_COLLECTION] in serial mode
INFO  : Starting task [Stage-0:MOVE] in serial mode
INFO  : Loading data to table default.test from hdfs://slave2.hadoop:8020/wareho
use/tablespace/managed/hive/test/.hive-staging_hive_2019-09-25_14-41-21_074_6410
419709428797468-1/-ext-10000
INFO  : Starting task [Stage-3:STATS] in serial mode
INFO  : Completed executing command(queryId=hive_20190925144121_14690c77-e122-4a
e7-92f1-5f7533dbe7f6); Time taken: 8.974 seconds
INFO  : OK
No rows affected (9.557 seconds)
```

图 5-10 删除 test 表中的数据

删除之后，进行查询操作，如图 5-11 所示。

test.id	test.name	test.subject	test.grade
1	Tony	Chinese	89
3	Mark	Physics	85
4	Grace	Math	86

图 5-11 删除操作后的查询结果

④ 使用 where 语句对 test 表进行查询操作，如图 5-12 所示。

```
0: jdbc:hive2://slave2.hadoop:2181,slave4.had> select * from test where  name='M
ark';
INFO  : Compiling command(queryId=hive_20190925145426_6ab36c28-8708-4315-96d1-02
18811d0856): select * from test where  name='Mark'
INFO  : Semantic Analysis Completed (retrial = false)
INFO  : Returning Hive schema: Schema(fieldSchemas:[FieldSchema(name:test.id, ty
pe:int, comment:null), FieldSchema(name:test.name, type:string, comment:null), F
ieldSchema(name:test.subject, type:string, comment:null), FieldSchema(name:test.
grade, type:int, comment:null)], properties:null)
INFO  : Completed compiling command(queryId=hive_20190925145426_6ab36c28-8708-43
15-96d1-0218811d0856); Time taken: 0.31 seconds
INFO  : Executing command(queryId=hive_20190925145426_6ab36c28-8708-4315-96d1-02
18811d0856): select * from test where  name='Mark'
INFO  : Completed executing command(queryId=hive_20190925145426_6ab36c28-8708-43
15-96d1-0218811d0856); Time taken: 0.012 seconds
INFO  : OK
+----------+------------+---------------+-------------+
| test.id  | test.name  | test.subject  | test.grade  |
+----------+------------+---------------+-------------+
| 3        | Mark       | Physics       | 85          |
+----------+------------+---------------+-------------+
```

图 5-12　使用 where 语句的查询结果

⑤ 排序。根据 age 对 test 表的数据进行降序排列，操作如图 5-13 所示。

```
0: jdbc:hive2://slave2.hadoop:2181,slave4.had> select * from test order by age d
esc;

+----------+------------+---------------+-------------+
| test.id  | test.name  | test.subject  | test.grade  |
+----------+------------+---------------+-------------+
| 1        | Tony       | Chinese       | 89          |
| 4        | Grace      | Math          | 86          |
| 3        | Mark       | Physics       | 85          |
+----------+------------+---------------+-------------+
```

图 5-13　使用排序语句的查询结果

7. 参考答案

Hive 分布式数据仓库的架构和工作原理如下所述。

图 5-14 显示了 Hive 的主要模块以及 Hive 是如何与 Hadoop 交互工作的。

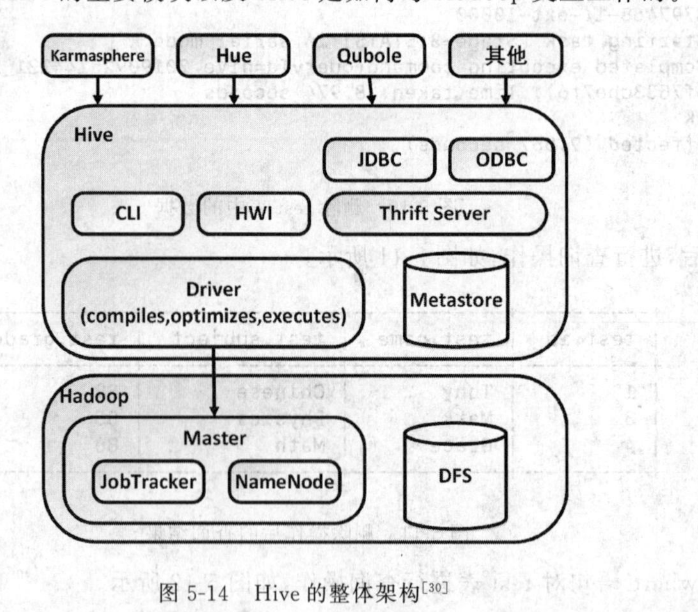

图 5-14　Hive 的整体架构[30]

所有的命令和查询都会进入Driver（驱动模块），通过该模块对输入进行解析编译，对需求的计算进行优化，然后按照指定的步骤执行〔通常是启动多个MapReduce任务(job)来执行〕。当需要启动MapReduce任务时，Hive本身是不会生成Java MapReduce算法程序的。相反地，Hive通过一个表示“job执行计划”的XML文件驱动执行内置的、原生的Mapper和Reducer模块。换句话说，这些通用的模块函数类似于微型的语言翻译程序，而这个驱动计算的“语言”是以XML形式编码的[30]。

Hive通过和JobTracker通信来初始化MapReduce任务，而不必部署在JobTracker所在的管理节点上执行。在大型集群中，通常会有网关机专门用于部署类似于Hive的工具，在这些网关机上可远程和管理节点上的JobTracker通信来执行任务。通常，要处理的数据文件是存储在HDFS中的，而HDFS是由NameNode进行管理的。Metastore（元数据存储）是一个独立的关系型数据体（通常是一个MySQL实例），Hive会在其中保存表模式和其他系统元数据[30]。

实验作业②、③的答案见本节扩展资料部分。

5.2 Druid的安装

1. 实验目的

学会安装分布式查询引擎Druid。

2. 实验要求

① Java 8。

② Linux、MacOSX或其他类似于Unix的操作系统（不支持Windows）。

③ 16 GB内存。

④ 4核CPU。

3. 预备知识

Druid是一款支持数据实时写入、低延时、高性能的联机分析处理(OLAP)引擎，具有优秀的数据聚合能力与实时查询能力，在大数据分析、实时计算、监控等领域都有特定的应用场景，是大数据基础架构建设中重要的一环[31]。

Druid是针对时间序列数据提供低延时数据写入以及快速交互式查询的分布式OLAP数据库，其两大关键点是：首先，Druid主要针对时间序列数据提供低延时数据写入和快速聚合查询；其次，Druid是一款分布式OLAP引擎[31]。

4. 实验内容

① 下载apache-druid-0.16.0-incubating安装包。

② 在终端运行以下命令来提取Druid。

```
tar -xzf apache-druid-0.16.0-incubating-bin.tar.gz
cd apache-druid-0.16.0-incubating
```

③ Druid依赖于Apache ZooKeeper的分布式协调，因此需要下载运行ZooKeeper。在Druid包的根目录中，运行以下命令。

```
curl https://archive.apache.org/dist/zookeeper/zookeeper-3.4.11/zookeeper-3.4.11.tar.gz -o zookeeper-3.4.11.tar.gz
tar -xzf zookeeper-3.4.11.tar.gz
mv zookeeper-3.4.11 zk
```

④ 在 apache-druid-0.16.0-incubating 包的根目录下，运行以下命令。

```
./bin/start-micro-quickstart
```

以下是本地运行 ZooKeeper 和 Druid 服务的结果。

```
bin/supervise -c quickstart/tutorial/conf/tutorial-cluster.conf
[Wed Sep 25 17:02:44 2019] Running command[coordinator-overlord], logging to[/root/druid/apache-druid-0.16.0-incubating/var/sv/coordinator-overlord.log]: bin/run-druid coordinator-overlord conf/druid/single-server/micro-quickstart
[Wed Sep 25 17:02:44 2019] Running command[broker], logging to[/root/druid/apache-druid-0.16.0-incubating/var/sv/broker.log]: bin/run-druid broker conf/druid/single-server/micro-quickstart
[Wed Sep 25 17:02:44 2019] Running command[router], logging to[/root/druid/apache-druid-0.16.0-incubating/var/sv/router.log]: bin/run-druid router conf/druid/single-server/micro-quickstart
[Wed Sep 25 17:02:44 2019] Running command[historical], logging to[/root/druid/apache-druid-0.16.0-incubating/var/sv/historical.log]: bin/run-druid historical conf/druid/single-server/micro-quickstart
[Wed Sep 25 17:02:44 2019] Running command[middleManager], logging to[/root/druid/apache-druid-0.16.0-incubating/var/sv/middleManager.log]: bin/run-druid middleManager conf/druid/single-server/micro-quickstart
```

⑤ Druid 服务的日志位于 apache-druid-0.16.0-incubating 包的根目录下的 var/sv 中。如果要停止服务，可通过快捷键“Ctrl+C”退出 bin/start-micro-quickstart 脚本，终止 Druid 进程。如果想重新设置集群状态，将服务停止后，请删除 var 目录并运行 bin/start-micro-quickstart 再写一遍。

5. 实验作业

① 找到 Druid 安装包，在计算机上安装分布式查询引擎 Druid。

② 熟悉 Druid 的整体架构，了解 Druid 的内部组件，包括 Historical、MiddleManager、Broker、Coordinator、Overlord、Router 等核心进程和 Deep Storage、Metadata Store、ZooKeeper 三个外部依赖项。

6. 扩展资料

① 下载 Druid 安装包到本地 Linux 操作系统计算机，如图 5-15 所示。

```
[root@slave2 druid]# ls
apache-druid-0.16.0-incubating-bin.tar.gz
[root@slave2 druid]# 
```

图 5-15　下载 Druid 安装包

② 运行命令提取 Druid，如图 5-16 所示。

```
[root@slave2 druid]# tar -xzf apache-druid-0.16.0-incubating-bin.tar.gz
[root@slave2 druid]# cd apache-druid-0.16.0-incubating
[root@slave2 apache-druid-0.16.0-incubating]# ls
bin   DISCLAIMER  hadoop-dependencies  LICENSE   NOTICE      README
conf  extensions  lib                  licenses  quickstart
[root@slave2 apache-druid-0.16.0-incubating]#
```

图 5-16　解压 Druid 安装包

③ 从 https://archive-apache. org/dist/zookeeper/zookeeper-3. 4. 14/zookeeper-3. 4. 14. tar. gz 下载 ZooKeeper 到 Druid 包的根目录并解压，如图 5-17 所示。

```
[root@slave2 apache-druid-0.16.0-incubating]# tar -xzf zookeeper-3.4.14.tar.gz
[root@slave2 apache-druid-0.16.0-incubating]# ls
bin         hadoop-dependencies  NOTICE            zookeeper-3.4.14.tar.gz
conf        lib                  quickstart
DISCLAIMER  LICENSE              README
extensions  licenses             zookeeper-3.4.14
[root@slave2 apache-druid-0.16.0-incubating]# mv zookeeper-3.4.14 zk
[root@slave2 apache-druid-0.16.0-incubating]# ls
bin         extensions           LICENSE   quickstart  zookeeper-3.4.14.tar.gz
conf        hadoop-dependencies  licenses  README
DISCLAIMER  lib                  NOTICE    zk
```

图 5-17　解压 ZooKeeper

④ 在 Druid 包的根目录下运行启动命令，启动成功的结果如图 5-18 所示。

```
[root@slave2 apache-druid-0.16.0-incubating]# ./bin/start-micro-quickstart
[Wed Sep 25 17:02:44 2019] Running command[zk], logging to[/root/druid/apache-dr
uid-0.16.0-incubating/var/sv/zk.log]: bin/run-zk conf
[Wed Sep 25 17:02:44 2019] Running command[coordinator-overlord], logging to[/ro
ot/druid/apache-druid-0.16.0-incubating/var/sv/coordinator-overlord.log]: bin/ru
n-druid coordinator-overlord conf/druid/single-server/micro-quickstart
[Wed Sep 25 17:02:44 2019] Running command[broker], logging to[/root/druid/apach
e-druid-0.16.0-incubating/var/sv/broker.log]: bin/run-druid broker conf/druid/si
ngle-server/micro-quickstart
[Wed Sep 25 17:02:44 2019] Running command[router], logging to[/root/druid/apach
e-druid-0.16.0-incubating/var/sv/router.log]: bin/run-druid router conf/druid/si
ngle-server/micro-quickstart
[Wed Sep 25 17:02:44 2019] Running command[historical], logging to[/root/druid/a
pache-druid-0.16.0-incubating/var/sv/historical.log]: bin/run-druid historical c
onf/druid/single-server/micro-quickstart
[Wed Sep 25 17:02:44 2019] Running command[middleManager], logging to[/root/drui
d/apache-druid-0.16.0-incubating/var/sv/middleManager.log]: bin/run-druid middle
Manager conf/druid/single-server/micro-quickstart
```

图 5-18　运行 Druid

⑤ 按“Ctrl+C”即可关闭 Druid 服务，结果如图 5-19 所示。

```
^C[Wed Sep 25 17:04:52 2019] Sending signal[15] to command[middleManager] (timeo
ut 10s).
[Wed Sep 25 17:04:56 2019] Command[middleManager] exited (pid = 18439, exited =
143)
[Wed Sep 25 17:04:56 2019] Sending signal[15] to command[coordinator-overlord] (
timeout 10s).
[Wed Sep 25 17:04:56 2019] Sending signal[15] to command[broker] (timeout 10s).
[Wed Sep 25 17:04:56 2019] Sending signal[15] to command[router] (timeout 10s).
[Wed Sep 25 17:04:56 2019] Sending signal[15] to command[historical] (timeout 10
s).
[Wed Sep 25 17:04:57 2019] Command[coordinator-overlord] exited (pid = 18435, ex
ited = 143)
[Wed Sep 25 17:04:58 2019] Command[broker] exited (pid = 18436, exited = 143)
[Wed Sep 25 17:04:58 2019] Command[router] exited (pid = 18437, exited = 143)
[Wed Sep 25 17:04:58 2019] Command[historical] exited (pid = 18438, exited = 143
)
[Wed Sep 25 17:04:58 2019] Sending signal[15] to command[zk] (timeout 10s).
[Wed Sep 25 17:04:58 2019] Command[zk] exited (pid = 18434, exited = 143)
[Wed Sep 25 17:04:58 2019] Exiting.
```

图 5-19　关闭 Druid

7. 参考答案

实验作业①的答案见本节扩展资料部分。

实验作业②的答案如下所述，下面介绍 Druid 的核心组件和外部依赖项。

Historical(历史节点)进程是处理存储和查询“历史”数据的主要工具。Historical 进程从深层存储中下载 Segment 并响应与这些 Segment 有关的查询，不接受写数据。

MiddleManager(中间管理者节点)进程处理对新数据的摄入，负责从外部数据源读取数据，并发布新的 Segment 数据文件。

Broker(查询节点)进程从外部客户端接收查询，并将这些查询转发给 Historicals 和 MiddleManagers。当 Brokers 从这些子查询中收到结果时，它们会合并这些结果并将其返回给调用者。最终用户通常会查询 Brokers，而不是直接查询 Historicals 或 MiddleManagers。

Coordinator(协调节点)进程监视 Historical 进程，负责将 Segment 分配给特定服务器，并确保 Segment 在 Historical 进程之间保持平衡。

Overlord(统治节点)进程监视 MiddleManager 进程，并且是数据读入 Druid 的控制器，负责将摄取任务分配给 MiddleManagers 并协调 Segment 发布。

Router(路由节点)进程是可选的进程，它在 Druid Brokers、Overlords 和 Coordinators 之前提供统一的 API 网关。Router 进程是可选的，也可以直接联系 Druid Brokers、Overlords 和 Coordinators。

Deep Storage(深度存储)：每个 Druid 服务器都可以访问共享文件存储，通常是类似于 S3 或 HDFS 的分布式对象存储，或者是网络安装的文件系统，Druid 使用 Deep Storage 来存储系统中已经读取的任何数据。

Metadata Store(元数据库)：共享元数据存储，这通常是一个传统的关系数据库管理系统(RDBMS)，如 PostgreSQL 或 MySQL。

ZooKeeper(分布式协调服务)用于内部服务发现、协调和领导人选举。

Druid 的整体架构如图 5-20 所示。

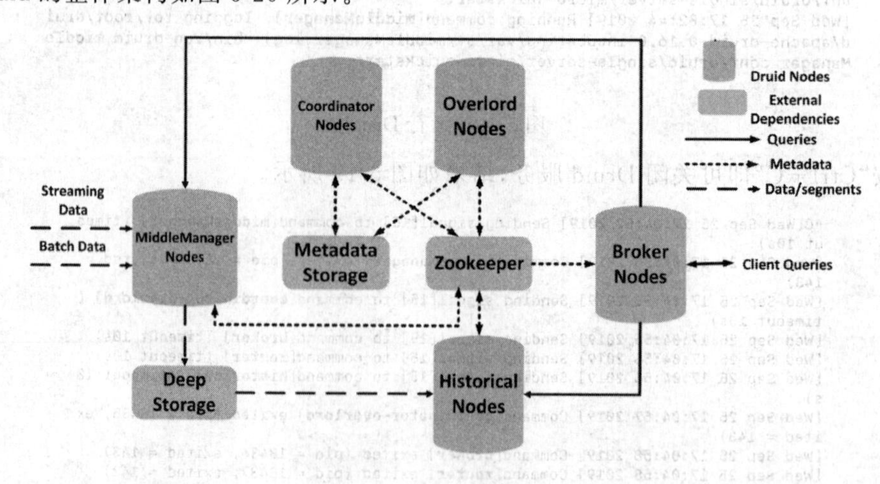

图 5-20　Druid 的整体架构

5.3 Druid的数据摄入与数据查询

1. 实验目的

① 学会使用分布式查询引擎 Druid 进行数据摄入。

② 学会使用 Python 编写客户端对 Druid 中的数据进行查询。

2. 实验要求

① 安装了分布式查询引擎 Druid 的计算机。

② Python 3。

③ Python 编译器，推荐使用 PyCharm。

3. 预备知识

Druid 支持通过 JSON-over-HTTP 和 SQL 查询数据。Druid 包含多种查询类型，如对用户摄入 Druid 的数据进行 TopN、Timeseries、GroupBy、Select、Search 等方式的查询，也可以查询一个数据源的 timeBoundary、segmentMetadata、dataSourceMetadata 等。[32]

PyDruid 是 Python 里的一个包，它公开了一个简单的 API 来创建、执行和分析 Druid 查询。PyDruid 可以将查询结果解析为 Pandas 的 DataFrame 对象，用于进行后续的数据分析。[33]

4. 实验内容

① 数据加载是通过提交摄入任务给 Druid 的 Overlord 进程。为方便起见，已配置为读取 quickstart/tutorial/wikiticker-2015-09-12-sampled.json.gz 文件。若要读取 HDFS 中的文件，需要修改数据源的配置，ioConfig 指明了真正的、具体的数据源。Druid 包括一个批处理摄取脚本，该脚本位于 bin/post-index-task，它向 Overlord 进程提交一个摄取任务，并轮询 Druid，直到数据可用于查询为止。[34]

首先，启动 Druid 服务，然后在 Druid 包的根目录下运行以下命令。

```
bin/post-index-task --file quickstart/tutorial/wikipedia-index.json --url http://localhost:8081
```

输出结果如下所示。

```
Beginning indexing data for wikipedia
Task started: index_wikipedia_2019-09-25T09:22:44.195Z
Task log: http://localhost:8081/druid/indexer/v1/task/index_wikipedia_2019-09-25T09:22:44.195Z/log
Task status: http://localhost:8081/druid/indexer/v1/task/index_wikipedia_2019-09-25T09:22:44.195Z/status
Task index_wikipedia_2019-09-25T09:22:44.195Z still running...
Task index_wikipedia_2019-09-25T09:22:44.195Z still running...
Task index_wikipedia_2019-09-25T09:22:44.195Z still running...
Task finished with status: SUCCESS
```

```
Completed indexing data for wikipedia. Now loading indexed data onto the cluster...
wikipedia is 0.0 % finished loading...
wikipedia loading complete! You may now query your data
```

② 数据查询是通过使用 Python 中的 PyDruid 包来查询 Druid 中的数据。

a. 在 PyCharm 中建立 Python 工程。

拓展阅读

Druid 查询

b. 将 PyDruid 包加入 Python 工程中。首先,在项目设置处(settings)打开 Python 解释器,打开后可以看到项目中包的情况,然后单击添加按钮"+"进行 PyDruid 包的添加。

c. 编写 Druid 的查询客户端,对 Wikipedia 数据源进行 TopN、GroupBy、Timeseries 等方式的查询。

以下是 TopN 查询的示例代码。

```
# _*_ coding: utf-8 _*_
from pydruid.client import PyDruid
from pydruid.utils.aggregators import count
from pydruid.utils.filters import Dimension

query = PyDruid('http://安装 druid 的计算机 IP:8082', 'druid/v2/')
langs = query.topn(
    datasource = "wikipedia",
    granularity = "all",
    intervals = "2015-09-12/2015-09-13",
    dimension = "page",
    aggregations = {"count": count("count")},
    metric = "count",
    threshold = 5
)
for value in langs[0]['result']:
    print (value)
```

TopN 查询的结果如下所示。

```
{'count': 33, 'page': 'Wikipedia:Vandalismusmeldung'},
{'count': 28, 'page': 'User:Cyde/List of candidates for speedy deletion/Subpage'},
{'count': 27, 'page': 'Jeremy Corbyn'},
{'count': 21, 'page': "Wikipedia:Administrators' noticeboard/Incidents"},
{'count': 20, 'page': 'Flavia Pennetta'}
```

进行 GroupBy 查询时,需要将 langs 变量和结果输出替换,如下所示。

```
langs = query.groupby(
        datasource = "wikipedia",
        granularity = "day",
        intervals = "2015-09-12/2015-09-13",
        dimensions = ["page","cityName"],
        aggregations = {"count": count("count")},
        limit_spec = {"limit":"10",
        "type":"default",
        "columns":[
        {"dimension":"count","direction":"descending"}
        ]
        }
    )
for value in langs:
print(value['event'])
```

GroupBy 查询的结果如下所示。

```
{'page': 'Wikipedia:Vandalismusmeldung', 'count': 31, 'cityName': None}
{'page': 'User:Cyde/List of candidates for speedy deletion/Subpage', 'count': 28,
 'cityName': None}
{'page': 'Jeremy Corbyn', 'count': 25, 'cityName': None}
{'page': "Wikipedia: Administrators' noticeboard/Incidents", 'count': 19,
 'cityName': None}
{'page': 'User talk:Dudeperson176123', 'count': 18, 'cityName': None}
{'page': 'Wikipédia:Le Bistro/12 septembre 2015', 'count': 18, 'cityName': None}
{'page': 'Wikipedia:Requests for page protection', 'count': 17, 'cityName': None}
{'page': 'Wikipedia:Administrator intervention against vandalism', 'count': 16,
 'cityName': None}
{'page': 'Utente:Giulio Mainardi/Sandbox', 'count': 16, 'cityName': None}
{'page': 'Wikipedia:In the news/Candidates', 'count': 15, 'cityName': None}
```

进行 Timeseries 查询时,需要将 langs 变量和结果输出替换,如下所示。

```
langs = query.timeseries(
     datasource = "wikipedia",
     granularity = "hour",
     intervals = "2015-09-12/2015-09-13",
     aggregations = {"count": count("count")},
     filter = Dimension('page') == 'Wikipedia:Vandalismusmeldung'
)
for value in langs:
 print(value)
```

Timeseries 查询的结果如下所示。

```
{'result': {'count': 0}, 'timestamp': '2015-09-12T00:00:00.000Z'}
{'result': {'count': 0}, 'timestamp': '2015-09-12T01:00:00.000Z'}
{'result': {'count': 0}, 'timestamp': '2015-09-12T02:00:00.000Z'}
{'result': {'count': 0}, 'timestamp': '2015-09-12T03:00:00.000Z'}
{'result': {'count': 2}, 'timestamp': '2015-09-12T04:00:00.000Z'}
{'result': {'count': 0}, 'timestamp': '2015-09-12T05:00:00.000Z'}
{'result': {'count': 1}, 'timestamp': '2015-09-12T06:00:00.000Z'}
{'result': {'count': 0}, 'timestamp': '2015-09-12T07:00:00.000Z'}
{'result': {'count': 3}, 'timestamp': '2015-09-12T08:00:00.000Z'}
{'result': {'count': 1}, 'timestamp': '2015-09-12T09:00:00.000Z'}
{'result': {'count': 9}, 'timestamp': '2015-09-12T10:00:00.000Z'}
{'result': {'count': 3}, 'timestamp': '2015-09-12T11:00:00.000Z'}
{'result': {'count': 0}, 'timestamp': '2015-09-12T12:00:00.000Z'}
{'result': {'count': 0}, 'timestamp': '2015-09-12T13:00:00.000Z'}
{'result': {'count': 1}, 'timestamp': '2015-09-12T14:00:00.000Z'}
{'result': {'count': 0}, 'timestamp': '2015-09-12T15:00:00.000Z'}
{'result': {'count': 1}, 'timestamp': '2015-09-12T16:00:00.000Z'}
{'result': {'count': 1}, 'timestamp': '2015-09-12T17:00:00.000Z'}
{'result': {'count': 3}, 'timestamp': '2015-09-12T18:00:00.000Z'}
{'result': {'count': 2}, 'timestamp': '2015-09-12T19:00:00.000Z'}
{'result': {'count': 2}, 'timestamp': '2015-09-12T20:00:00.000Z'}
{'result': {'count': 3}, 'timestamp': '2015-09-12T21:00:00.000Z'}
{'result': {'count': 0}, 'timestamp': '2015-09-12T22:00:00.000Z'}
{'result': {'count': 1}, 'timestamp': '2015-09-12T23:00:00.000Z'}
```

5. 实验作业

① 按照实验内容的步骤,完成 Wikipedia 数据源的导入。

② 利用 Python 编写 Druid 的客户端,对 Wikipedia 数据源完成 TopN、GroupBy、Timeseries 等方式的查询。

6. 扩展资料

① 完成 Wikipedia 数据源的导入。

a. 启动 Druid 服务,启动成功结果如图 5-21 所示。

```
[root@slave2 apache-druid-0.16.0-incubating]# ./bin/start-micro-quickstart
[Wed Sep 25 17:02:44 2019] Running command[zk], logging to[/root/druid/apache-dr
uid-0.16.0-incubating/var/sv/zk.log]: bin/run-zk conf
[Wed Sep 25 17:02:44 2019] Running command[coordinator-overlord], logging to[/ro
ot/druid/apache-druid-0.16.0-incubating/var/sv/coordinator-overlord.log]: bin/ru
n-druid coordinator-overlord conf/druid/single-server/micro-quickstart
[Wed Sep 25 17:02:44 2019] Running command[broker], logging to[/root/druid/apach
e-druid-0.16.0-incubating/var/sv/broker.log]: bin/run-druid broker conf/druid/si
ngle-server/micro-quickstart
[Wed Sep 25 17:02:44 2019] Running command[router], logging to[/root/druid/apach
e-druid-0.16.0-incubating/var/sv/router.log]: bin/run-druid router conf/druid/si
ngle-server/micro-quickstart
[Wed Sep 25 17:02:44 2019] Running command[historical], logging to[/root/druid/a
pache-druid-0.16.0-incubating/var/sv/historical.log]: bin/run-druid historical c
onf/druid/single-server/micro-quickstart
[Wed Sep 25 17:02:44 2019] Running command[middleManager], logging to[/root/drui
d/apache-druid-0.16.0-incubating/var/sv/middleManager.log]: bin/run-druid middle
Manager conf/druid/single-server/micro-quickstart
```

图 5-21 启动 Druid 服务

b. 运行命令导入 Wikipedia 数据源，导入成功的结果如图 5-22 所示。

```
[root@slave2 apache-druid-0.16.0-incubating]# bin/post-index-task --file quickst
art/tutorial/wikipedia-index.json --url http://localhost:8081
Beginning indexing data for wikipedia
Task started: index_wikipedia_2019-09-25T09:22:44.195Z
Task log:     http://localhost:8081/druid/indexer/v1/task/index_wikipedia_2019-0
9-25T09:22:44.195Z/log
Task status:  http://localhost:8081/druid/indexer/v1/task/index_wikipedia_2019-0
9-25T09:22:44.195Z/status
Task index_wikipedia_2019-09-25T09:22:44.195Z still running...
Task index_wikipedia_2019-09-25T09:22:44.195Z still running...
Task index_wikipedia_2019-09-25T09:22:44.195Z still running...
Task index_wikipedia_2019-09-25T09:22:44.195Z still running...
Task index_wikipedia_2019-09-25T09:22:44.195Z still running...
Task index_wikipedia_2019-09-25T09:22:44.195Z still running...
Task index_wikipedia_2019-09-25T09:22:44.195Z still running...
Task index_wikipedia_2019-09-25T09:22:44.195Z still running...
Task index_wikipedia_2019-09-25T09:22:44.195Z still running...
Task index_wikipedia_2019-09-25T09:22:44.195Z still running...
Task finished with status: SUCCESS
Completed indexing data for wikipedia. Now loading indexed data onto the cluster
...
wikipedia is 0.0% finished loading...
wikipedia loading complete! You may now query your data
```

图 5-22　导入 Wikipedia 数据源

② 利用 Python 编写 Druid 的客户端，对 Wikipedia 数据源完成 TopN、GroupBy、Timeseries 等方式的查询。

a. 在 PyCharm 中建立 Python 工程，如图 5-23 所示。

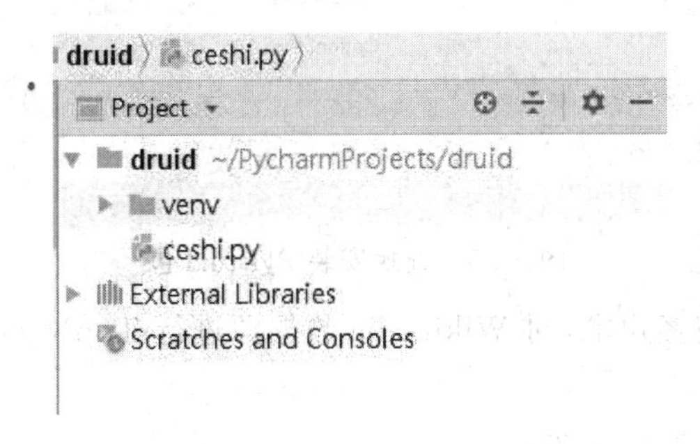

图 5-23　建立 Python 工程

b. 将 PyDruid 包加入 Python 工程。

首先，在项目设置(settings)处打开 Python 解释器(Interpreter)，打开后可以看到项目中包的情况，如图 5-24 所示。

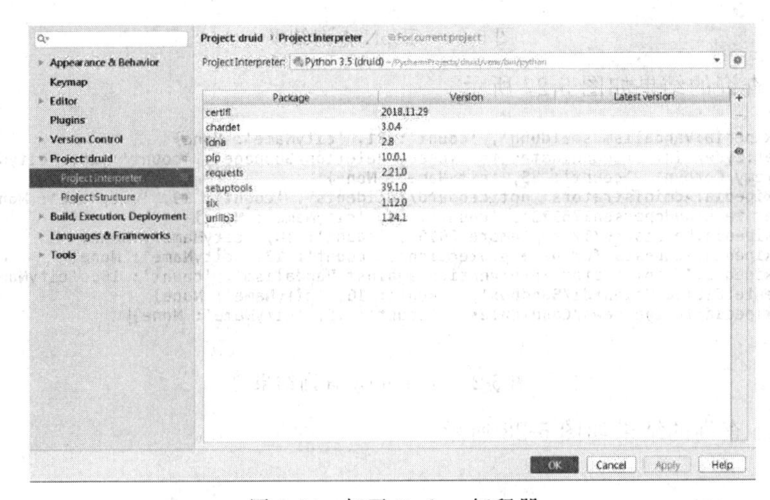

图 5-24　打开 Python 解释器

然后单击包的添加按钮“+”，会弹出包的搜索窗口，输入“pydruid”进行查找，找到PyDruid包后进行安装，安装成功后的结果如图5-25所示。

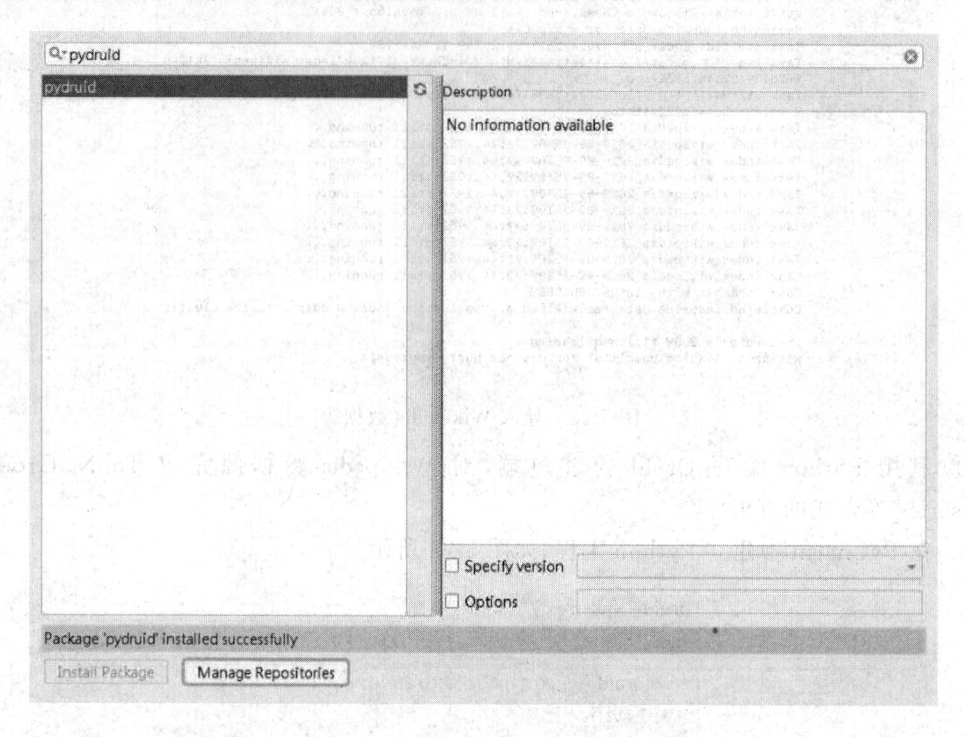

图5-25　查找安装PyDruid包

c. 编写Druid的查询客户端，对Wikipedia数据源进行TopN、GroupBy、Timeseries等方式的查询。

TopN查询的结果如图5-26所示。

```
{'count': 33, 'page': 'Wikipedia:Vandalismusmeldung'},
{'count': 28, 'page': 'User:Cyde/List of candidates for speedy deletion/Subpage'},
{'count': 27, 'page': 'Jeremy Corbyn'},
{'count': 21, 'page': "Wikipedia:Administrators' noticeboard/Incidents"},
{'count': 20, 'page': 'Flavia Pennetta'}]
```

图5-26　TopN查询结果

GroupBy查询的结果如图5-27所示。

```
{'page': 'Wikipedia:Vandalismusmeldung', 'count': 31, 'cityName': None}
{'page': 'User:Cyde/List of candidates for speedy deletion/Subpage', 'count': 28, 'cityName': None}
{'page': 'Jeremy Corbyn', 'count': 25, 'cityName': None}
{'page': "Wikipedia:Administrators' noticeboard/Incidents", 'count': 19, 'cityName': None}
{'page': 'User talk:Dudeperson176123', 'count': 18, 'cityName': None}
{'page': 'Wikipédia:Le Bistro/12 septembre 2015', 'count': 18, 'cityName': None}
{'page': 'Wikipedia:Requests for page protection', 'count': 17, 'cityName': None}
{'page': 'Wikipedia:Administrator intervention against vandalism', 'count': 16, 'cityName': None}
{'page': 'Utente:Giulio Mainardi/Sandbox', 'count': 16, 'cityName': None}
{'page': 'Wikipedia:In the news/Candidates', 'count': 15, 'cityName': None}
```

图5-27　GroupBy查询结果

Timeseries查询的结果如图5-28所示。

```
{'result': {'count': 0}, 'timestamp': '2015-09-12T00:00:00.000Z'}
{'result': {'count': 0}, 'timestamp': '2015-09-12T01:00:00.000Z'}
{'result': {'count': 0}, 'timestamp': '2015-09-12T02:00:00.000Z'}
{'result': {'count': 0}, 'timestamp': '2015-09-12T03:00:00.000Z'}
{'result': {'count': 2}, 'timestamp': '2015-09-12T04:00:00.000Z'}
{'result': {'count': 0}, 'timestamp': '2015-09-12T05:00:00.000Z'}
{'result': {'count': 1}, 'timestamp': '2015-09-12T06:00:00.000Z'}
{'result': {'count': 0}, 'timestamp': '2015-09-12T07:00:00.000Z'}
{'result': {'count': 3}, 'timestamp': '2015-09-12T08:00:00.000Z'}
{'result': {'count': 1}, 'timestamp': '2015-09-12T09:00:00.000Z'}
{'result': {'count': 9}, 'timestamp': '2015-09-12T10:00:00.000Z'}
{'result': {'count': 3}, 'timestamp': '2015-09-12T11:00:00.000Z'}
{'result': {'count': 0}, 'timestamp': '2015-09-12T12:00:00.000Z'}
{'result': {'count': 0}, 'timestamp': '2015-09-12T13:00:00.000Z'}
{'result': {'count': 1}, 'timestamp': '2015-09-12T14:00:00.000Z'}
```

图 5-28 Timeseries 查询结果

7. 参考答案

实验作业的答案见本节扩展资料部分。

5.4 Drill 的部署

1. 实验目的

在单机中基于嵌入模式安装 Apache Drill，并保证可以进入 Drill 命令行，使其正常运行。

2. 实验要求

① Java 8。

② Linux 操作系统。

③ 8 GB 内存(可更改配置进行调节)。

3. 预备知识

Drill 是一个低延迟的分布式海量数据(涵盖结构化、半结构化以及嵌套数据)交互式查询引擎，使用 ANSI SQL 兼容语法，支持本地文件、HDFS、Hive、HBase、MongoDB 等后端存储，支持 Parquet、JSON、CSV、TSV、PSV 等数据格式。

Drill 是 Google Dremel 的开源实现，其本质是一个分布式的大规模并行处理(MPP)查询层，支持 SQL 及一些用于 NoSQL 和 Hadoop 数据存储系统的语言，有助于 Hadoop 用户更快地查询海量数据集。

4. 实验内容

① 下载 Drill 的官方安装包。

```
wget http://apache.mirrors.hoobly.com/drill/drill-1.15.0/apache-drill-1.15.0.tar.gz
```

② 移动安装包到指定的安装目录下，本实验安装到/usr/local 目录下，同时解压到另一目录下。

```
mv apache-drill-1.15.0.tar.gz /usr/local/
cd /usr/local
tar -xvzf apache-drill-1.15.0.tar.gz
mkdir drill
mv apache-drill-1.15.0/* drill/
rm -rf apache-drill-1.15.0*
```

③ 在 Drill 安装目录的 bin 目录下开启嵌入式的 Drill 命令行。

```
cd /usr/local/drill/bin
./drill-embedded
```

以下是运行命令行后的正常运行结果。

```
Apache Drill 1.15.0
"Let's Drill something more solid than concrete."
0: jdbc:drill:zk = local >
```

④ 进入 Drill 命令行后，其 Web 端界面也会默认开启在 localhost:8047 端口下，可以输入相应网址进行查看。

5. 实验作业

按照实验步骤，完成 Drill 的部署，保证进入分布式的命令行后可以正常访问网页。

6. 扩展资料

① 下载 Drill 的官方压缩包后，结果如图 5-29 所示。

```
[root@slave2 drill]# ls
apache-drill-1.16.0.tar.gz
```

图 5-29 目录下的 Drill 压缩包

② 移动并解压安装包，同时删除原压缩包，如图 5-30 和图 5-31 所示。

```
[root@slave2 drill]# mv apache-drill-1.16.0.tar.gz /usr/local/
[root@slave2 drill]# cd /usr/local/
[root@slave2 local]# tar -xzvf apache-
apache-drill-1.16.0.tar.gz          apache-kylin-2.4.1-bin-hbase1x.tar.gz
[root@slave2 local]# tar -xzvf apache-drill-1.16.0.tar.gz
```

图 5-30 移动并解压安装包

```
[root@node1 local]# mkdir drill
[root@node1 local]# mv apache-drill-1.15.0/* drill/
[root@node1 local]# rm -rf apache-drill-1.15.0*
[root@node1 local]# ls
bin  drill  etc  games  include  lib  lib64  libexec  sbin  share  src
```

图 5-31 规整文件目录

③ 进入 Drill 的命令行，如图 5-32 所示。

```
[root@slave2 bin]# ./drill-embedded
Apache Drill 1.16.0
"A Drill in the hand is better than two in the bush."
apache drill>
```

图 5-32 打开 Drill 命令行

④ 访问相应的界面，如图 5-33 所示。

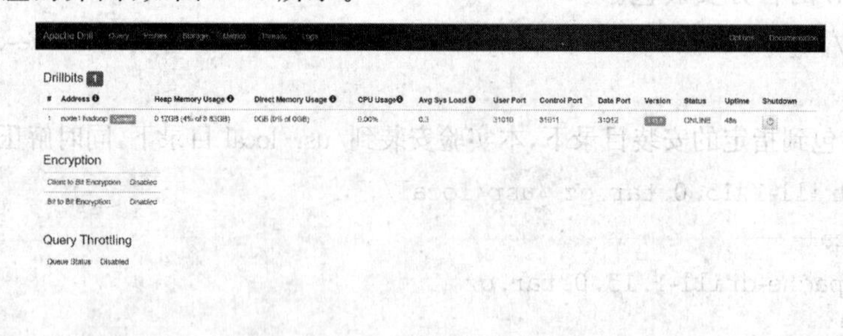

图 5-33 Drill 前端界面

7. 参考答案

实验作业的答案见本节扩展资料部分。

5.5 Drill 命令行与 PyDrill 的基础使用

1. 实验目的

在命令行中通过执行 SQL 语句进行文件的读取，使用 pydrill 库进行基础的文件读取操作。

2. 实验要求

① Python 2/3。

② pip 环境。

③ 已下载的 Drill 相关文件。

3. 预备知识

在启动 drill-embedded 之后，可以使用 SQL 语句进行各类文件的读取，本实验用到的示例文件 employee.json 打包在 Drill 第三方库中，目录为./jars/3rdparty/foodmart-data-json.0.4.jar，可以直接进行文件的读取。

PyDrill 为 Apache Drill 的 Python 驱动程序，支持修改配置、读取文件等多种操作，本实验主要用于读取文件。[35]

4. 实验内容

① 在 Drill 的命令行中输入以下命令，可以看到示例文件中的部分数据。[36]

```
SELECT * FROM cp.'employee.json' LIMIT 5;
```

② 打开一个新的终端，保持 Drill 为开启状态，然后准备使用 Python 进行 Drill 相关开发，首先安装所需的依赖库 pydrill(如果安装速度较慢，建议切换为国内的 pip 源)。

```
pip install pydrill
```

③ PyDrill 包成功安装后，可以使用 Python 代码进行开发，示例代码如下所示，其中的路径应设置为主机中实际的文件路径。

```
from pydrill.client import PyDrill
drill = PyDrill(host = 'localhost', port = 8047)

if not drill.is_active():
    print("Please start it first")
data = drill.query('''
  SELECT * FROM `dfs.root`.`./root/employee.json`
  LIMIT 5
''')
df = data.to_dataframe()
print df
```

④ 运行该 Python 文件后可以得到以下结果，可以看到文件前五行的信息。

```
    birth_date department_id    education_level employee_id  ...          position_
title    salary store_id supervisor_id
    0  1961-08-26              1   Graduate Degree           1  ...
President   80000.0        0            0
    1  1915-07-03              1   Graduate Degree           2  ...       VP Country
Manager   40000.0        0            1
    2  1969-06-20              1   Graduate Degree           4  ...       VP Country
Manager   40000.0        0            1
    3  1951-05-10              1   Bachelors Degree          5  ...       VP Country
Manager   35000.0        0            1
    4  1942-10-08              2   Bachelors Degree          6  ...   VP Information
Systems   25000.0        0            1

    [5 rows x 16 columns]
```

5. 实验作业

在可以正常执行 SQL 的基础上，安装 PyDrill 依赖，实现使用 Python 连接 Drill 并读取指定文件的功能。

拓展阅读

Drill 连接 Hive 数据源进行查询

6. 扩展资料

① 在 Drill 的命令行中输入指定的命令后，结果如图 5-34 所示。

```
apache drill> select * from cp.`employee.json` limit 2;
+-------------+----------------+------------+-----------+-------------+------
---------+----------+---------------+------------+-----------------------+---
---------+---------------+-----------------+----------------+--------+-------
-------+
| employee_id |   full_name    | first_name | last_name | position_id |   posit
ion_title   | store_id | department_id | birth_date |       hire_date       | sa
lary  | supervisor_id | education_level | marital_status | gender |  management_
role  |
+-------------+----------------+------------+-----------+-------------+------
---------+----------+---------------+------------+-----------------------+---
---------+---------------+-----------------+----------------+--------+-------
-------+
| 1           | Sheri Nowmer   | Sheri      | Nowmer    | 1           | Preside
nt          | 0        | 1             | 1961-08-26 | 1994-12-01 00:00:00.0 | 80
000.0 | 0             | Graduate Degree | S              | F      | Senior Manag
ement |
| 2           | Derrick Whelply | Derrick   | Whelply   | 2           | VP Coun
try Manager | 0        | 1             | 1915-07-03 | 1994-12-01 00:00:00.0 | 40
000.0 | 1             | Graduate Degree | M              | M      | Senior Manag
ement |
+-------------+----------------+------------+-----------+-------------+------
---------+----------+---------------+------------+-----------------------+---
---------+---------------+-----------------+----------------+--------+-------
-------+
2 rows selected (0.559 seconds)
apache drill>
```

图 5-34　Drill 命令行返回数据

② 在新的终端中，编辑一个 demo. py 文件，输入指定的代码，如图 5-35 所示。

```
from pydrill.client import PyDrill
drill = PyDrill(host='localhost', port=8047)

if not drill.is_active():
    print("Please start it first")

data = drill.query('''
  SELECT * FROM  `dfs.root`.`./home/fatbird/employee.json`
  LIMIT 5
''')
df = data.to_dataframe()
print df
```

图 5-35　Python 文件代码

③ 使用“python demo. py”运行该文件后，可以得到图 5-36 所示的输出结果。

```
[root@slave2 drill]# python demo.py
   birth_date department_id    ...      store_id supervisor_id
0  1961-08-26             1    ...             0             0
1  1915-07-03             1    ...             0             1
2  1969-06-20             1    ...             0             1
3  1951-05-10             1    ...             0             1
4  1942-10-08             2    ...             0             1

[5 rows x 16 columns]
[root@slave2 drill]# 
```

图 5-36 运行 Python 文件返回的数据

7. 参考答案

实验作业的答案见本节扩展资料部分。

第 6 章

大数据分析:Kylin 多维分析实验教程

6.1 Kylin 的安装

1. 实验目的

学会安装分布式分析引擎 Kylin。

2. 实验要求

(1) 软件要求

① Hadoop：2.7＋，3.1＋（自 v2.5 以后）。

② Hive：0.13～1.2.1＋。

③ HBase：1.1＋，2.0（自 v2.5 以后）。

④ Spark（可选）:2.1.1＋。

⑤ Kafka（可选）:0.10.0＋。

⑥ JDK：1.8＋（自 v2.5 以后）。

⑦ OS：Linux，CentOS 6.5＋或 Ubuntu 16.0.4＋。

(2) 硬件要求

运行 Kylin 的服务器的最低配置为 4 核 CPU、16 GB 内存和 100 GB 磁盘。对于高负载的场景,建议使用 24 核 CPU、64 GB 内存或更高的配置。

(3) Hadoop 环境

Kylin 依赖于 Hadoop 集群处理大量的数据集,因此需要准备一个配置好 HDFS、YARN、MapReduce、Hive、HBase、ZooKeeper 和其他服务的 Hadoop 集群供运行。

Kylin 可以在 Hadoop 集群的任意节点上启动,用户可以在 master 节点上运行 Kylin,但为了更好的稳定性,建议部署在一个干净的 Hadoop client 节点上,该节点上 Hive、HBase、HDFS 等命令行已安装好且 client 配置(如 core-site. xml、hive-site. xml、hbase-site. xml 等)也已经合理地配置,同时可以自动和其他节点同步。

运行的 Linux 账户要有访问 Hadoop 集群的权限,包括创建/写入 HDFS 文件夹、Hive 表、HBase 表和提交 MapReduce 任务的权限。

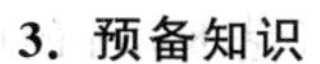

3. 预备知识

Kylin 是 eBay 开发的一套 OLAP 系统，主要用于支持大数据生态圈的数据分析业务，它主要是通过预计算的方式将用户设定的多维立方体缓存到 HBase 中。

通过 Kylin，用户可以与 Hadoop 数据进行亚秒级交互，在同样的数据集上提供比 Hive 更好的性能。

4. 实验内容

① 下载 Kylin 安装包并解压缩。

从 Apache Kylin 下载网站下载一个适用于 Hadoop 版本的二进制文件。例如，适用于 HBase 1.x 的 2.5.0 版本可通过如下命令行下载得到。

```
cd /usr/local/
wget http://mirror.bit.edu.cn/apache/kylin/apache-kylin-2.5.0/apache-kylin-2.5.0-bin-hbase1x.tar.gz
```

解压 tar 包，配置环境变量 $KYLIN_HOME 指向文件夹。

```
tar -zxvf apache-kylin-2.5.0-bin-hbase1x.tar.gz
cd apache-kylin-2.5.0-bin-hbase1x
export KYLIN_HOME=`pwd`
```

为了方便以后的访问，可以在用户家目录下的.bashrc 文件中加入：

```
export KYLIN_HOME= /usr/local/apache-kylin-2.5.0-bin-hbase1x
```

② 启动 Kylin 前检查相关权限和组件完整性。运行在 Hadoop 集群上，对各个组件的版本、访问权限及 classpath 等都有一定的要求，为了避免遇到各种环境问题，可以运行“$KYLIN_HOME/bin/check-env.sh”脚本来进行环境检测，如果环境存在问题，脚本将打印出详细报错信息，如果没有报错信息，代表环境适合运行。[37]

③ 启动 Kylin。运行“$KYLIN_HOME/bin/kylin.sh start”脚本来启动 Kylin，界面输出如下所示。

```
Retrieving hadoop conf dir...
KYLIN_HOME is set to /usr/local/apache-kylin-2.5.0-bin-hbase1x
......
A newinstance is started by root. To stop it, run 'kylin.sh stop'
Check the log at /usr/local/apache-kylin-2.5.0-bin-hbase1x/logs/kylin.log
Web UI is at http://<hostname>:7070/kylin
```

启动后可以查看 logs/kylin.log 日志信息，观察 Kylin 的启动过程，如果出现问题，请根据日志内容提示进行解决。

这里补充说明 Kylin 的日志文件，Kylin 启动后会创建 logs 目录，并在 logs 目录下产生以下几个日志文件。

- kylin.log 文件。该文件是主要的日志文件，所有的 logger 默认写入该文件，其中与 Kylin 相关的日志级别默认是 DEBUG。日志随日期轮转，即每天 0 时将前一天的日志存放到以日期为后缀的文件(如 kylin.log2016-08-06)中，并把新一天的日志保存到全新的 kylin.log 文件中。
- kylin.out 文件。该文件是标准输出的重定向文件，一些非 Kylin 产生的标准输出(如

tomcat 启动输出、Hive 命令行输出等)将被重定向到该文件。在执行完 Kylin 的启动命令后,查看 kylin.log 文件内容时会发现一段时间内并没有日志输出,这是因为此时正在启动 tomcat,日志都输入 kylin.out 文件中了。

- kylin.gc 文件。该文件用于记录 Kylin 的 Java 进程 GC 的日志内容。为避免多次启动覆盖旧文件,该日志使用进程号作为文件名后缀(如 kylin.gc.3862.0)。

从 Apache Kylin 1.5.3 版本开始,conf 目录下新增了 kylin-server-log4j.properties 文件。Kylin 使用 log4j 对日志进行配置,用户可以编辑 kylin-server-log4j.properties 文件,对日志级别、路径等进行修改,修改后,重新启动 Kylin 服务才可生效。

④ 通过 Web 访问 Kylin。Kylin 启动后可以通过在浏览器中输入"http://<hostname>:7070/kylin"进行访问,其中 <hostname> 为具体的机器名、IP 地址或域名,默认端口为 7070,登录页面如图 6-1 所示。

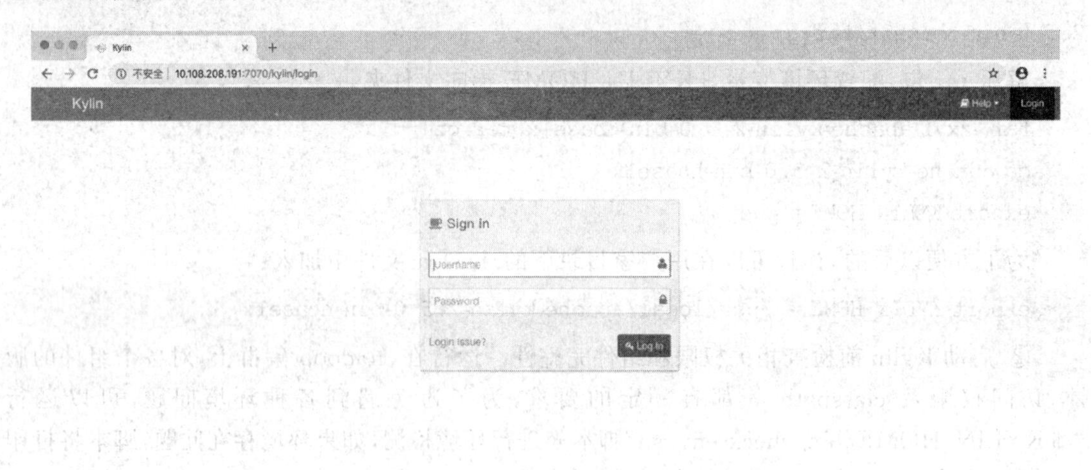

图 6-1 登录页面

初始用户名和密码是 ADMIN 和 KYLIN。服务器启动后,可以通过查看"$KYLIN_HOME/logs/kylin.log"获得运行时的日志。

⑤ 停止 Kylin。运行"$KYLIN_HOME/bin/kylin.sh stop"脚本来停止 Kylin,界面输出如下所示。

```
Retrieving hadoop conf dir...
KYLIN_HOME is set to /usr/local/apache-kylin-2.5.0-bin-hbase1x
Stopping Kylin: 25964
Stopping in progress. Will check after 2 secs again...
Kylin with pid 25964 has been stopped.
```

运行"ps -ef | grep"可以查看进程是否已停止。

6.2 Demo 案例实战

1. 实验目的

① 初步使用分布式分析引擎 Kylin，尤其是其 Web UI 页面。

② 通过使用 Kylin 自带的样例 Cube(多维立方体)，熟悉一个 Cube 从构建到查询的过程。

2. 实验要求

安装了分布式分析引擎 Kylin 的计算机。

3. 预备知识

Kylin 的思想是预计算，即对多维分析可能用到的度量进行预计算，将计算好的结果保存为 Cube 并存在 HBase 中，供查询时直接访问。将高复杂度的聚合运算、多表连接等操作转换成对预计算结果的查询，决定了 Kylin 能够拥有很好的快速查询和高并发能力。

因此，Cube 是 Kylin 的核心，本实验的目的是让读者对 Cube 建立一个初步的认识。本实验不对 Cube 的设计进行详细描述，Kylin 自带的样例 Cube 是已经设计好的，执行其基本操作即可。

4. 实验内容

① 准备数据。Kylin 准备了一个创建样例 Cube 脚本，脚本会创建 5 个样例 Hive 表，运行"$KYLIN_HOME/bin/sample.sh"脚本来导入这些数据。脚本执行完成后，需要刷新缓存，可以重启 Kylin 服务器或从 Web UI 页面加载元数据，本实验通过从 Web UI 加载元数据实现，如图 6-2 所示，单击 System 页面中最右侧"Actions"下的第一个选项"Reload Metadata"，即可完成元数据的重新加载。[38]

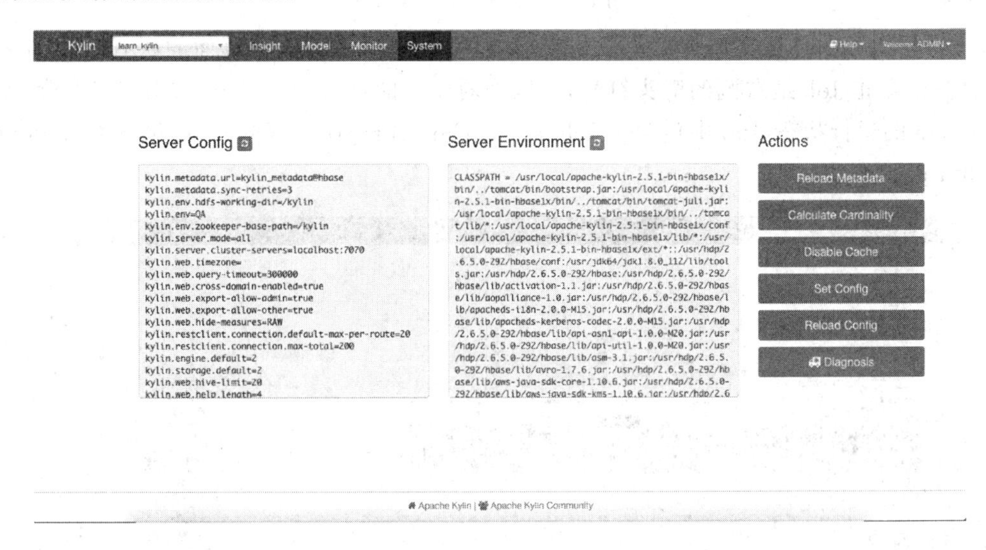

图 6-2　重新加载元数据

② 构建 Cube。

a. 选择 Project"learn_kylin"，在 Cubes 页面中单击 Cube 名称为"kylin_sales_cube"一栏右侧的"Actions"下拉框并选择"Build"操作，如图 6-3 所示。

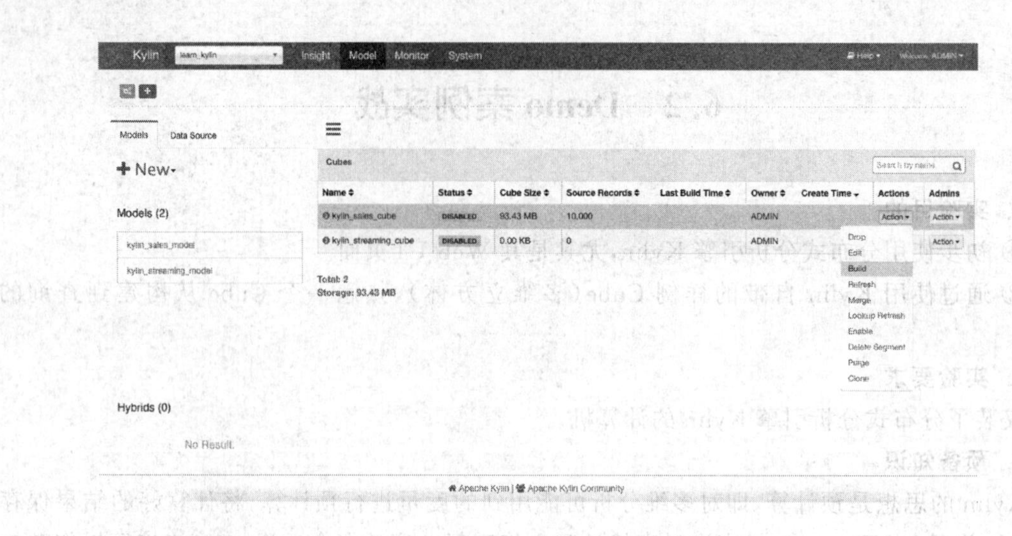

图 6-3 构建 Cube

b. 弹出窗口如图 6-4 所示，“End Date”处可任意选择。

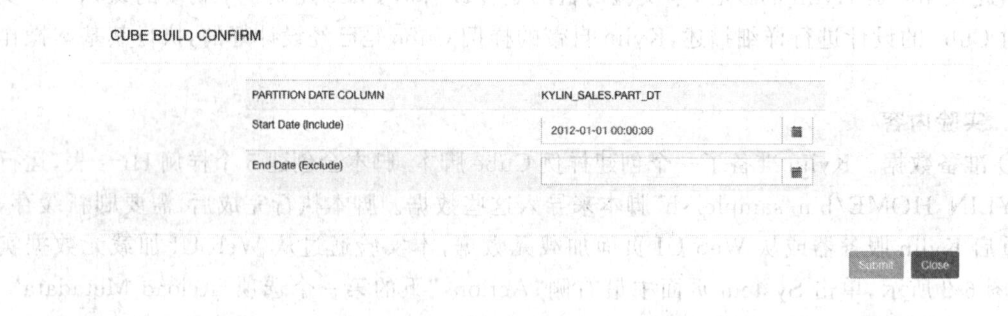

图 6-4 选择起止时间

c. 单击“Submit”提交请求。请求提交成功后，可以从 Monitor 页面的所有 Jobs 中找到新建的 Job，单击 Job 最右侧的箭头符号，可以查看 Job 的详细执行流程，如图 6-5 所示，其中标明了 Job 的运行状态，Job 中包含了 Job Name、Cube、Progress 等内容，还包含了 Actions 相关操作（Discard）。

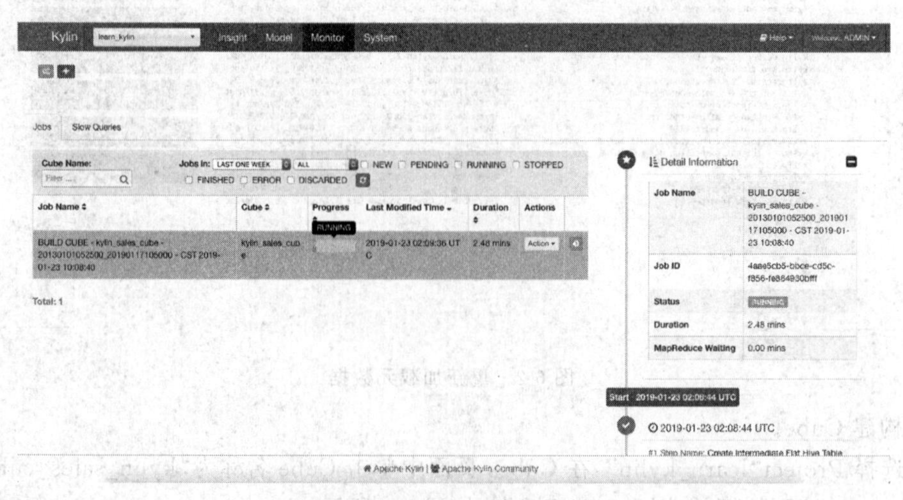

图 6-5 查看 Job 执行状态

d. Job 执行完成情况如图 6-6 所示。

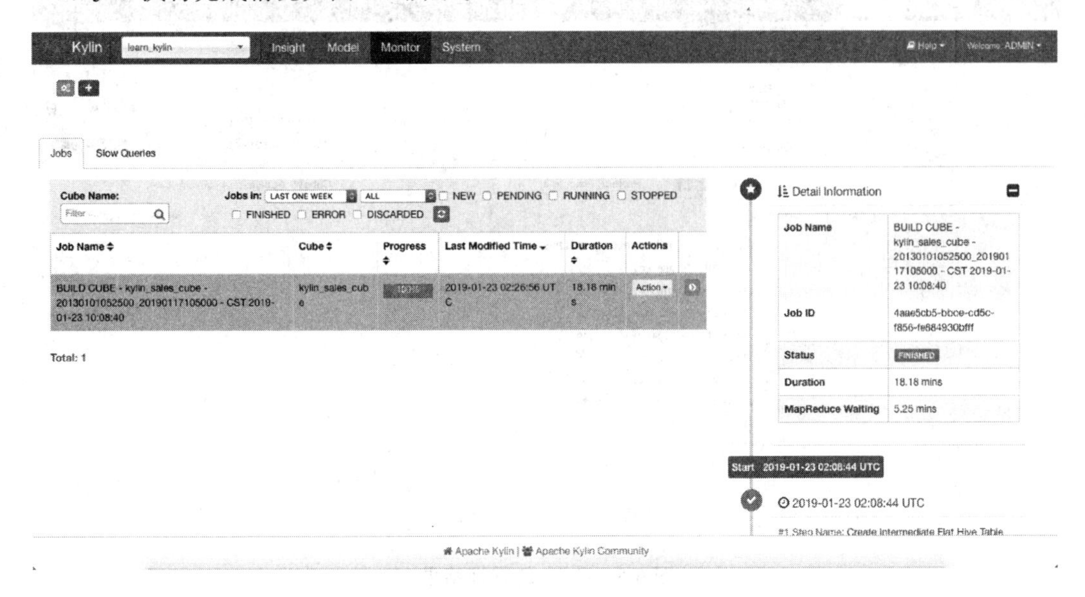

图 6-6　执行完成

③ 对构建好的 Cube 进行查询。切换到 Kylin 的 Insight 页面，其中提供交互式的 SQL 查询。

```
select part_dt, sum(price) as total selled, count(distinct seller_id) as sellers from kylin_sales group by part_dt order by part_dt;
```

将上述 SQL 语句写入 New Query 对话框中，然后单击右下角的“Submit”提交 SQL 查询，如图 6-7 所示。

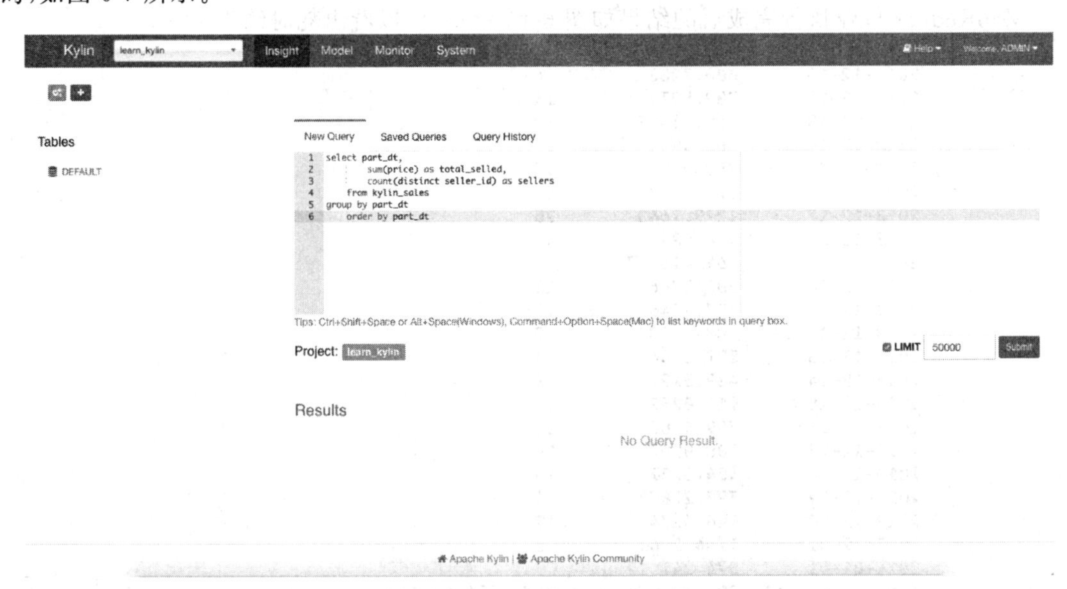

图 6-7　用 SQL 查询 Cube

提交后可以看到 0.18 s 后就返回了结果，如图 6-8 所示。

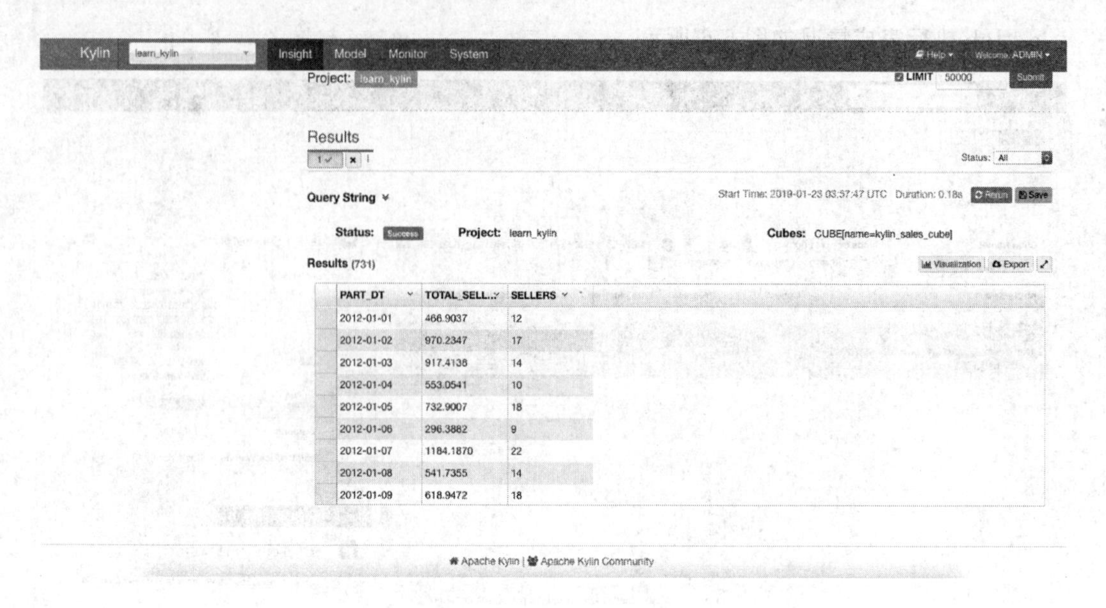

图 6-8 Cube 的查询结果

作为对比，再将上述 SQL 语句放到 Hive 内执行，如图 6-9 所示。

```
hive> select part_dt,
    >        sum(price) as total_selled,
    >        count(distinct seller_id) as sellers
    >     from kylin_sales
    > group by part_dt
    >     order by part_dt;
```

图 6-9 用 SQL 查询 Hive

MapReduce 作业执行完成后的结果如图 6-10 所示，可以看出耗时约为 11 s。

```
2013-12-11      600.7233        14
2013-12-12      709.5979        11
2013-12-13      1085.1017       19
2013-12-14      568.4304        9
2013-12-15      748.9417        17
2013-12-16      871.3873        16
2013-12-17      1279.7643       25
2013-12-18      874.9391        16
2013-12-19      762.416 17
2013-12-20      603.2746        13
2013-12-21      797.4346        17
2013-12-22      1004.7532       19
2013-12-23      587.1644        10
2013-12-24      409.5192        10
2013-12-25      912.0252        17
2013-12-26      799.8385        15
2013-12-27      708.0807        14
2013-12-28      434.5787        10
2013-12-29      797.2707        11
2013-12-30      926.5274        19
2013-12-31      1144.2961       18
2014-01-01      574.341 12
Time taken: 11.109 seconds, Fetched: 731 row(s)
hive>
```

图 6-10 Hive 的查询结果

通过比较可知，Kylin 的查询速度相对于 Hive 的有明显的提升。

6.3 多维分析的 Cube 创建实战

1. 实验目的

① 学会使用 Kylin 的 Web UI 页面设计一个 Model 和 Cube。

② 了解事实表、维度表(维表)、星型结构以及维度、度量等基本概念。

③ 了解 Build Cube 的步骤以及 Cube 的基本操作。

④ 基于自编数据，对生成的 Cube 进行查询和可视化处理。

2. 实验要求

① 安装了分布式分析引擎 Kylin 的计算机。

② Eclipse、IDEA 或其他 Java 编译器。

③ 命令行工具。

3. 预备知识

维度数据模型简称维度模型(DM，Dimensional Model)，是一套技术和概念的集合，用于数据仓库设计。

事实和维度是维度模型中的两个核心概念。事实表示对业务数据的度量，而维度是观察数据的角度。事实通常是数字类型的，可以进行聚合和计算，而维度通常是一组层次关系或描述信息，用于定义事实。例如，销售金额是一个事实，而销售时间、销售的产品、购买的顾客、商店等都是销售事实的维度。

维度模型通常以一种被称为星型模式的方式构建。所谓星型模式，就是以一个事实表为中心，周围环绕着多个维表。

事实表主要包含两方面的信息：维和度量。维的具体描述信息记录在维表中，事实表中的维属性只是一个关联到维表的键，并不记录具体信息；度量一般会记录事件的相应数值，如产品的购买数量、实付金额。维表中的信息一般是可以分层的，如时间维的年月日、地域维的省市县等，这类分层的信息是为了满足事实表中的度量可以在不同的粒度上完成聚合。[39]

拓展阅读

星型模式介绍

本实验将更深入地了解数据立方体 Cube 及其模型，并对其进行查询。

4. 实验内容

① Hive 中事实表以及多张多维表的处理。Kylin 处理的数据来源之一是 Hive，因此首先需要将分析的数据导入 Hive 中，并在 Hive 作为数据仓库层面进行预处理，目的是满足 Kylin 的 Cube 模型的要求。目前，迁移数据到 Hive 中的方式有很多，可以使用 ETL 工具(如 Kettle)或 Sqoop 开源组件。

本实验虽然比较简单，但是也会涵盖创建 Cube 的各个方面，本实验选择一张事实表和两张维表关联进行演示。

首先进入 Hive，建立并进入数据库。

```
hive> create database kylin_flat_db;
hive> use kylin_flat_db;
```

Hive 中事实表的建表语句如下所示。

```
create table kylin_flat_db.web_access_fact_tbl
(
    day date,
    cookieid string,
    regionid string,
    cityid string,
    siteid string,
    os string,
    pv bigint
)row format delimited
fields terminated by '|' stored as textfile;
```

事实表 web_access_fact_tbl 包含的维度有 day、regionid、cityid、siteid、os，度量指标包括汇总的 pv 和去重计数的 cookieid。

为了方便演示，本实验从本地加载数据到 Hive，数据文件 fact_data. txt 的部分内容如下所示。

```
2016-07-16|0ZPHGVNQB9BMRLFF21|G0501|9299|Android 5.0|9
2016-07-28|EN86ZQRYP461OCJIGQ|G0102|1264|Mac OS|2
2016-07-30|6IB6HVC3LLLM9NCPVQ|G0501|7640|Android 5.0|6
2016-07-22|TISP9C8KOQXO8R6OGV|G0502|7193|Mac OS|1
2016-07-22|N9VSKRJHKTHLOZVGN9|G0402|9010|Android 5.0|6
```

生成该文件的 Java 代码如下所示。

```
import java.io.BufferedWriter;
import java.io.File;
import java.io.FileWriter;
import java.io.IOException;
import java.util.Random;

public class CreateData {
    public static void main(String[] args){
        String cookieId[] = {"0", "1", "2", "3", "4", "5", "6", "7", "8", "9",
                             "Z", "Y", "X", "W", "V", "U", "T", "S", "R", "Q",
                             "P", "O", "N", "M", "L", "K", "J", "I", "H", "G",
                             "F", "E", "D", "C", "B", "A"};
        String regionId[] = {"G01", "G05", "G04", "G03", "G02"};
        String osId[] = {"Android 5.0", "Mac OS", "Windows 7"};
        String tempDate = null;
        String cookieIdTemp = null;
        String regionTemp = null;
        String cityTemp = null;
```

```
String sidTemp = null;
String osTemp = null;
String pvTemp = null;

try{
    for(int i = 0; i < 100; i++){
        int x = (int)(Math.random() * 31);
        if(x == 0){
            tempDate = "2016-07-01";
        }else if(x < 10){
            tempDate = "2016-07-0" + x;
        }else{
            tempDate = "2016-07-" + x;
        }

        for(int i1 = 0; i1 < 18; i1++){
            int j = (int)(Math.random() * 35);
            cookieIdTemp += cookieId[j];
        }

        int k = (int)(Math.random() * 4);
        regionTemp = regionId[k];

        int l = (int)(Math.random() * 2) + 1;
        cityTemp = regionTemp + "0" + l;

        Random random = new Random();
        int m = (int)Math.floor((random.nextDouble() * 10000.0));
        sidTemp = "" + m;

        int n = (int)(Math.random() * 2);
        osTemp = osId[n];

        int h = (int)(Math.random() * 9) + 1;
        pvTemp = "" + h;

        String data = tempDate + "|" + cookieIdTemp + "|" +
                  cityTemp + "|" + sidTemp + "|" + osTemp +
                  "|" + pvTemp + "\n";
```

```
                cookieIdTemp = "";

                File file = new File("fact_data.txt");

                if(!file.exists()){
                    file.createNewFile();
                }

                FileWriter fileWriter = new FileWriter(file.getName(), true);
                BufferedWriter bufferedWriter = new BufferedWriter(fileWriter);
                bufferedWriter.write(data);
                bufferedWriter.close();
            }
        }
        catch (IOException e){
            e.printStackTrace();
        }
    }
}
```

加载数据文件到 Hive 表中。

```
hive> use kyiln_flat_db;
hive> load data local inpath '/home/hdfs/fact_data.txt' into table_web_access_fact_tb1;
```

region 维度建表语句如下所示。

```
create table kylin_flat_db.region_tb1
(
    regionid string,
    regionname string
)row format delimited
fields terminated by '|' stored as textfile;
```

数据文件 region.txt 的内容如下所示。

```
G01|北京
G02|江苏
G03|浙江
G04|上海
G05|广东
```

city 维度建表语句如下所示。

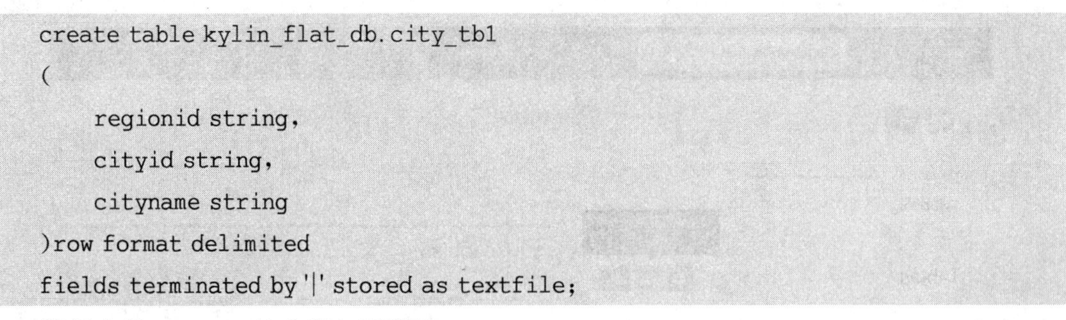

```
create table kylin_flat_db.city_tbl
(
    regionid string,
    cityid string,
    cityname string
)row format delimited
fields terminated by '|' stored as textfile;
```

数据文件 city. txt 的内容如下所述。

```
G01|G0101|朝阳
G01|G0102|海淀
G02|G0201|南京
G02|G0202|宿迁
G03|G0301|杭州
G03|G0302|嘉兴
G04|G0401|徐汇
G04|G0402|虹口
G05|G0501|广州
G05|G0502|珠海
```

将两张维表也加载到 Hive 表中。

```
hive> load data local inpath '/home/hdfs/region.txt' into table region_tbl;
hive> load data local inpath '/home/hdfs/city.txt' into table city_tbl;
```

② 在 Kylin 中建立项目、数据源。单击 Kylin Web 页面左上角的"+"，创建新项目，如图 6-11 所示。

New Project

Project Name

create_cube_test

Project Description

用于Kylin的学习及演示；

Project Config

+ Property

Close Submit

图 6-11 创建新项目

选择该项目，单击左侧的"Data Source"标签，有三种添加表格的方式，分别是"Load Table""Load Table From Tree""Add Streaming Table"，如图 6-12 所示。

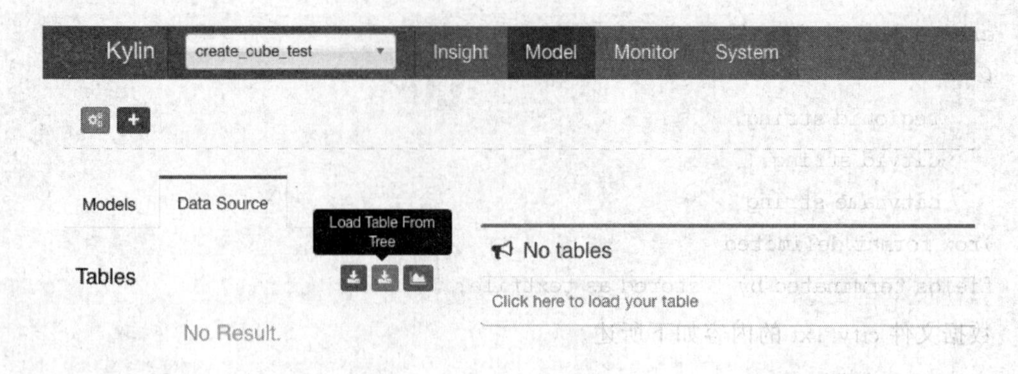

图 6-12　三种添加表格的方式

本实验选择“Load Table From Tree”，选择数据库“kylin_flat_db”下面的三张表，单击“Sync”进行同步，如图 6-13 所示。

图 6-13　添加表格

同步完成后界面如图 6-14 所示。

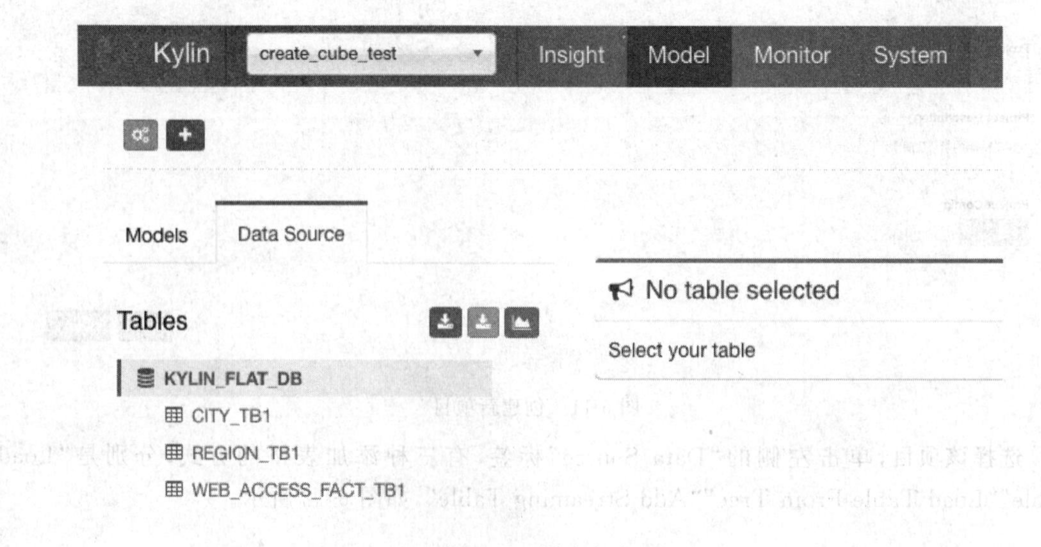

图 6-14　同步完成

③ 在 Kylin 中建立数据模型(Model)。在 Kylin 的 Web 页面中建立模型，单击 Models 页面的“New→New Model”，如图 6-15 所示。

图 6-15 建立数据模型

输入 Model 名称，如图 6-16 所示。

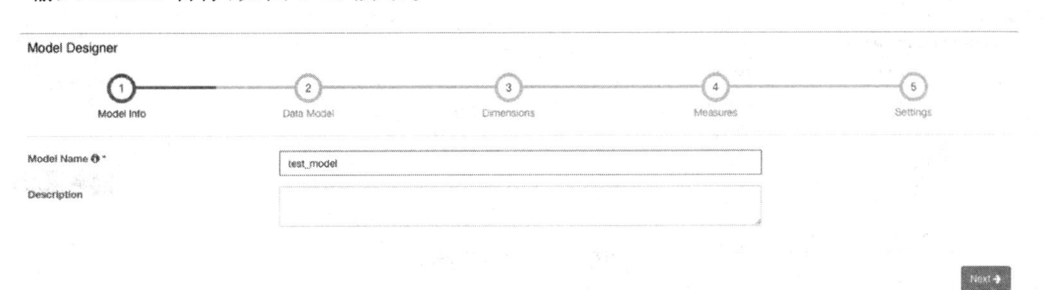

图 6-16 输入 Model 名称

单击“Next”，选择事实表和维表，单击“Add Lookup Table”添加维表，每次选择维表时，通过“left join”与事实表关联，具体设置如图 6-17 所示。

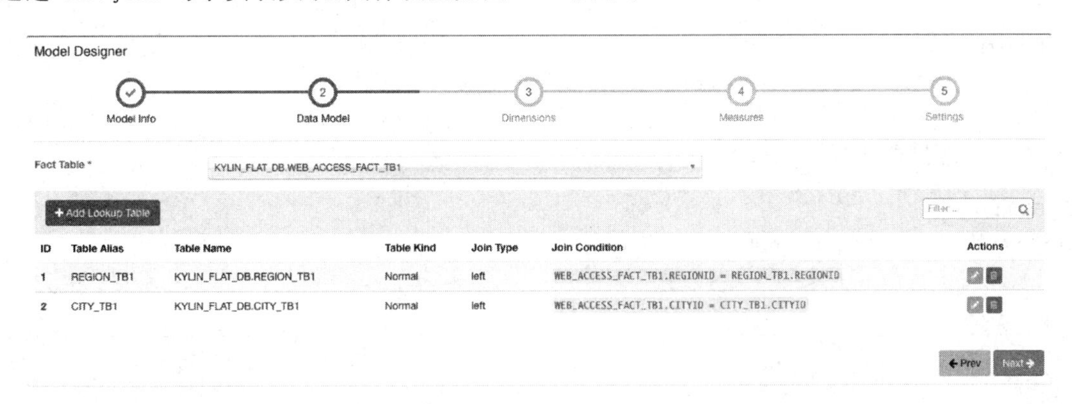

图 6-17 添加事实表和维表

单击“Next”,选择维度,如图 6-18 所示。

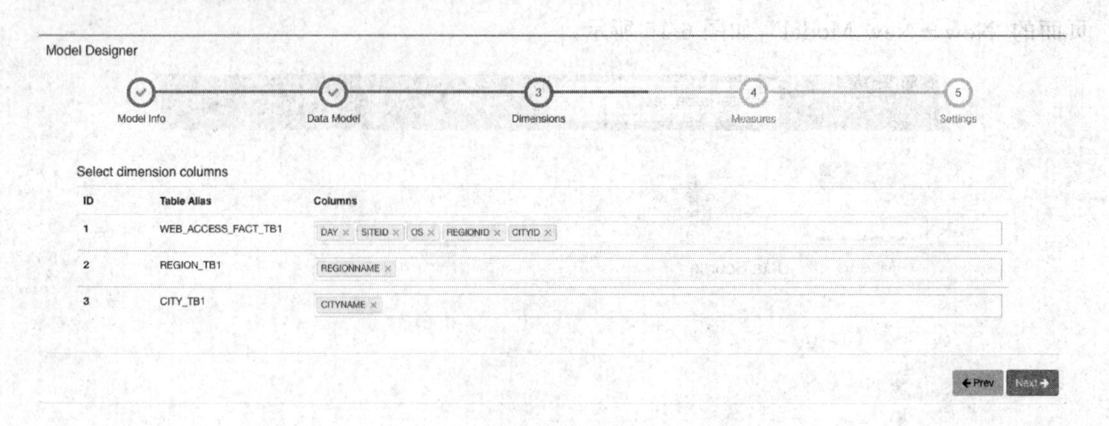

图 6-18　选择维度

选择好维度以后,单击“Next”,进一步选择度量,如图 6-19 所示。

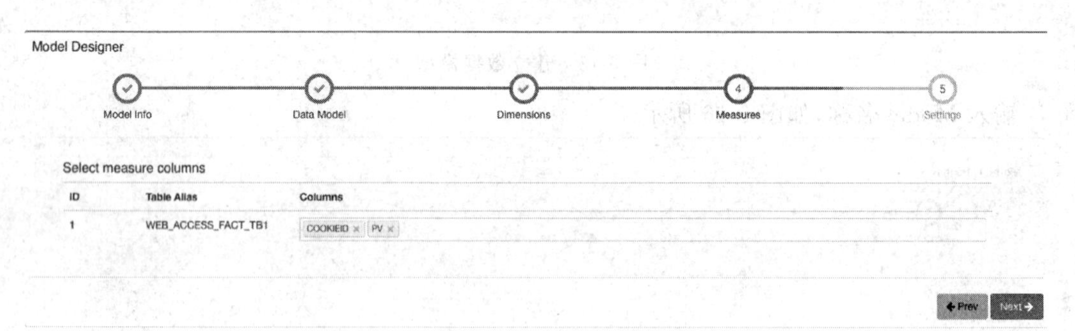

图 6-19　选择度量

最后,设置增量刷新字段,如图 6-20 所示。

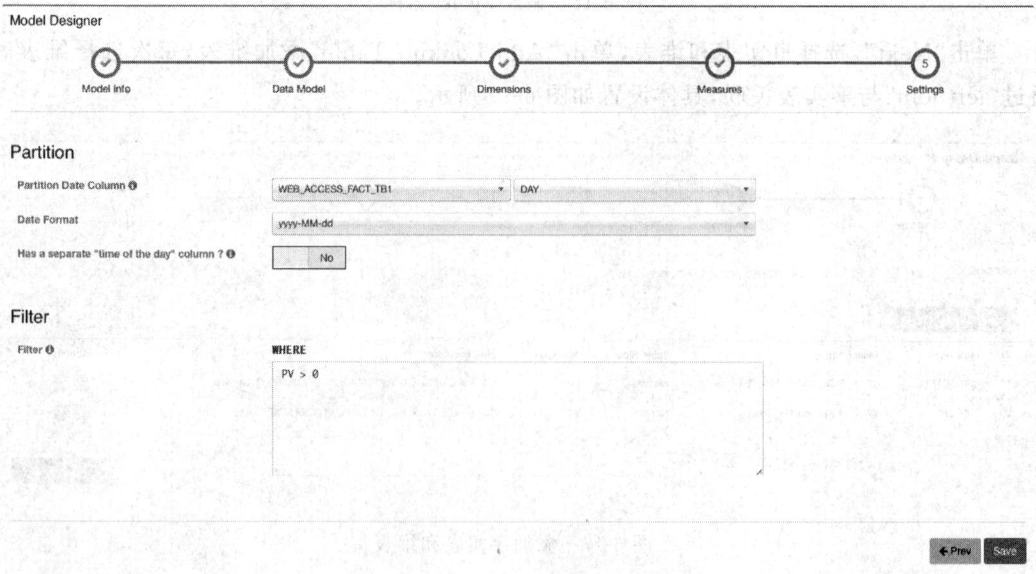

Model Designer

Model Info　Data Model　Dimensions　Measures　Settings

Partition

Partition Date Column　WEB_ACCESS_FACT_TB1　DAY

Date Format　yyyy-MM-dd

Has a separate "time of the day" column ?　No

Filter

Filter　WHERE

PV > 0

Prev　Save

图 6-20　设置增量刷新字段

模型创建完成以后，可以对模型进行检查，Kylin 提供了三种检查方式，分别是 Grid（表格）、Visualization（可视化）和 JSON（JSON 格式），可以单击 Models 页面中建立的 Model 的名称来查看。

a. Grid：创建 Model 的整个过程，如图 6-21 所示。

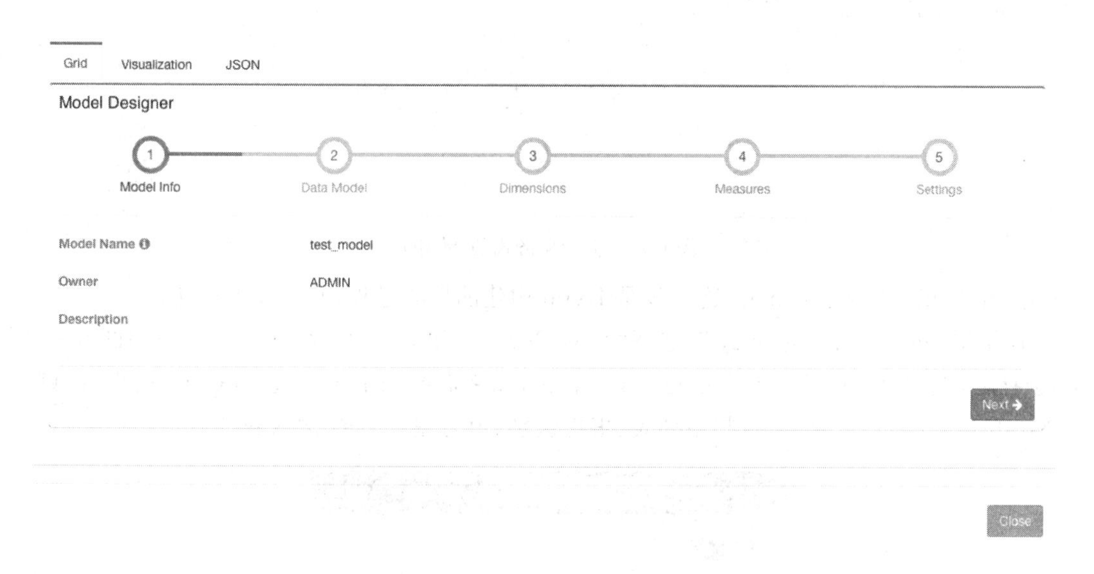

图 6-21 查看 Model 的创建过程

b. Visualization：可视化事实表和维表的关联，如图 6-22 所示。

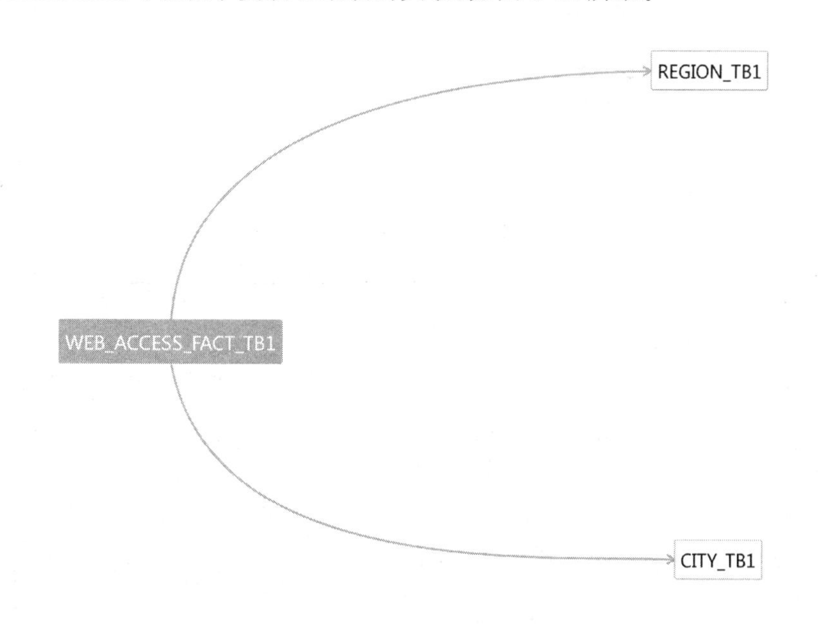

图 6-22 可视化事实表和维表的关联

c. JSON：以 JSON 格式配置 Model 模型中的表关联、维度字段、度量字段、Cube 刷新方式、过滤条件等，如图 6-23 所示。

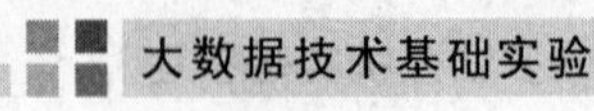

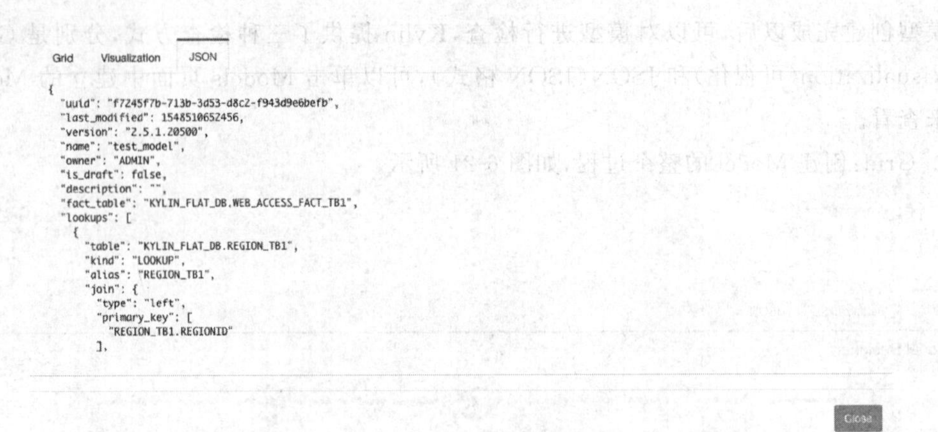

图 6-23　JSON 格式的 Model

④ 在 Kylin 中建立 Cube。这一步是 Kylin 构建的核心过程，操作如下所述。

单击 Models 页面下的“New”，选择“New Cube”如图 6-24 所示，弹出的界面如图 6-25 所示，选择之前创建的 Model，然后填写 Cube 名称，如果需要用 E-mail 通知 Cube 的事件，则填写 Notification 相关信息，包括邮箱列表、事件级别，填写完成后单击“Next”。

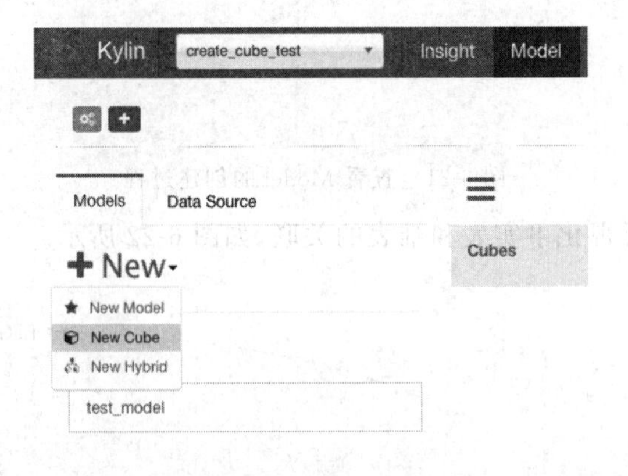

图 6-24　新建 Cube

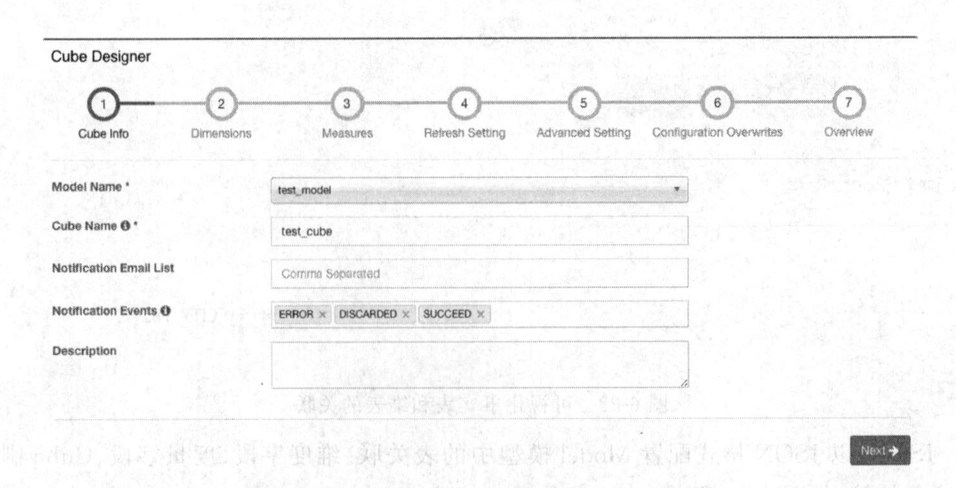

图 6-25　填写 Cube 基础信息

弹出图 6-26 所示的界面后，单击“Add Dimensions”，添加维度信息，可以看到维度有两种类型：normal 和 derived。normal 是最常见的类型，与所有其他的维度组合构成 Cuboid，derived 方式从有衍生维度(derived dimensions)的查找表来获取维度。

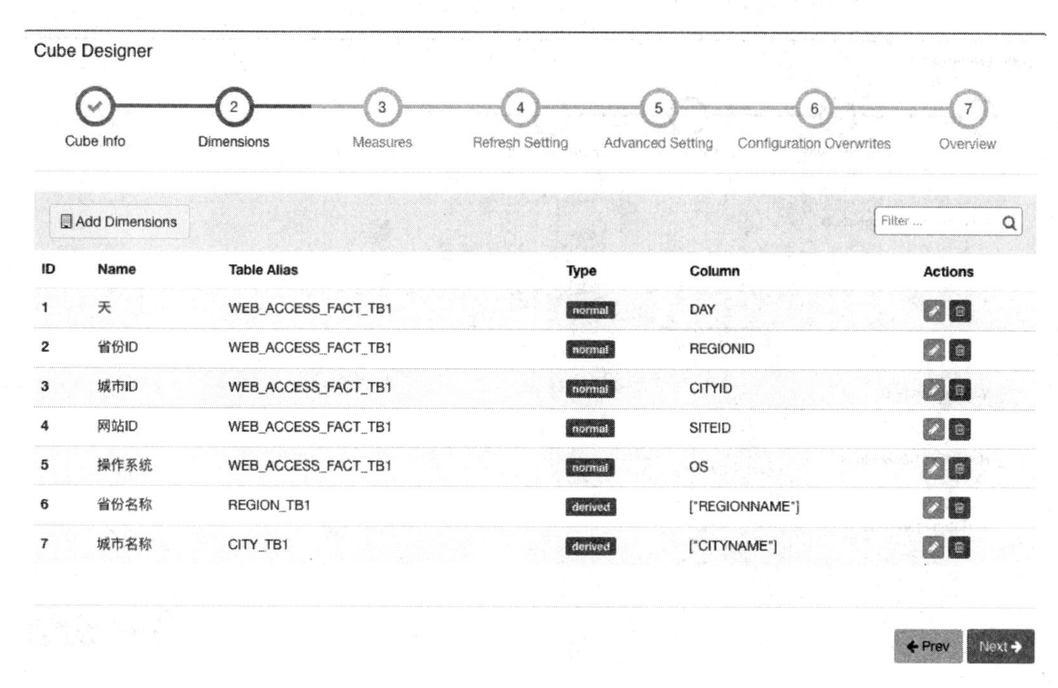

图 6-26　添加维度

添加好维度信息之后，单击“Next”，进入度量指标设计界面，如图 6-27 所示，可以通过“+Measure”图标进行度量值的增加，增加后的度量配置界面可以参考图 6-27，其中是对 pv 进行汇总计算(sum)，对 cookieid 进行去重过滤统计个数(count distinct)。

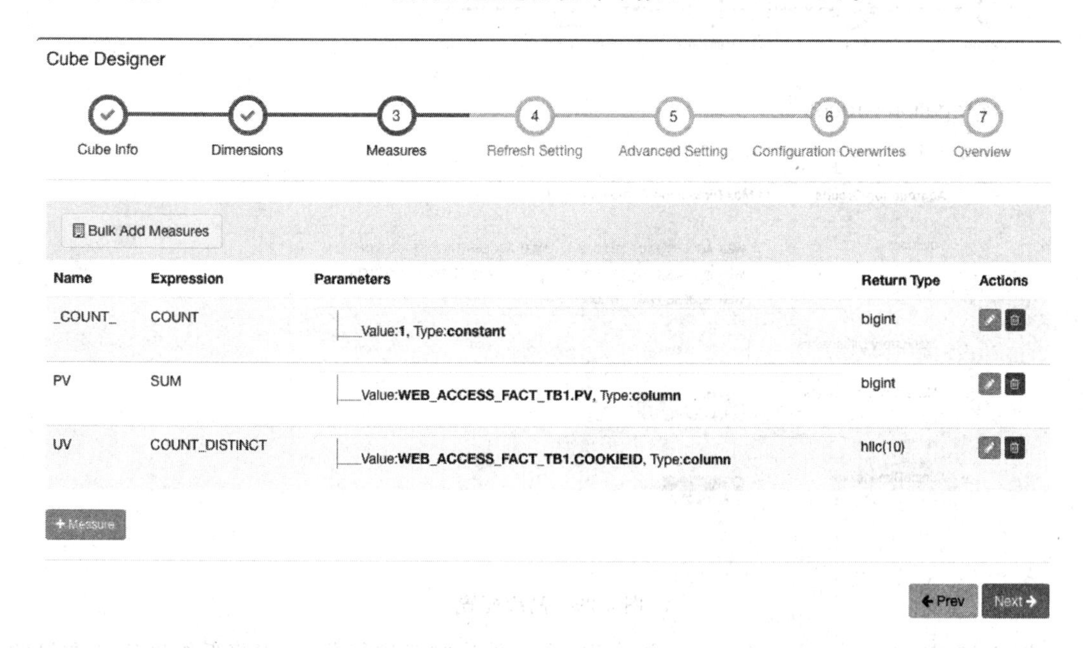

图 6-27　添加度量

继续单击"Next",进入 Cube 刷新界面设置,如图 6-28 所示,这部分主要用于设计增量 Cube 合并信息,默认每隔 7 天合并增量的 segments,每隔 28 天合并前面每 7 天合并的 segments,用户也可以自定义合并策略,单击"—"可以取消合并。

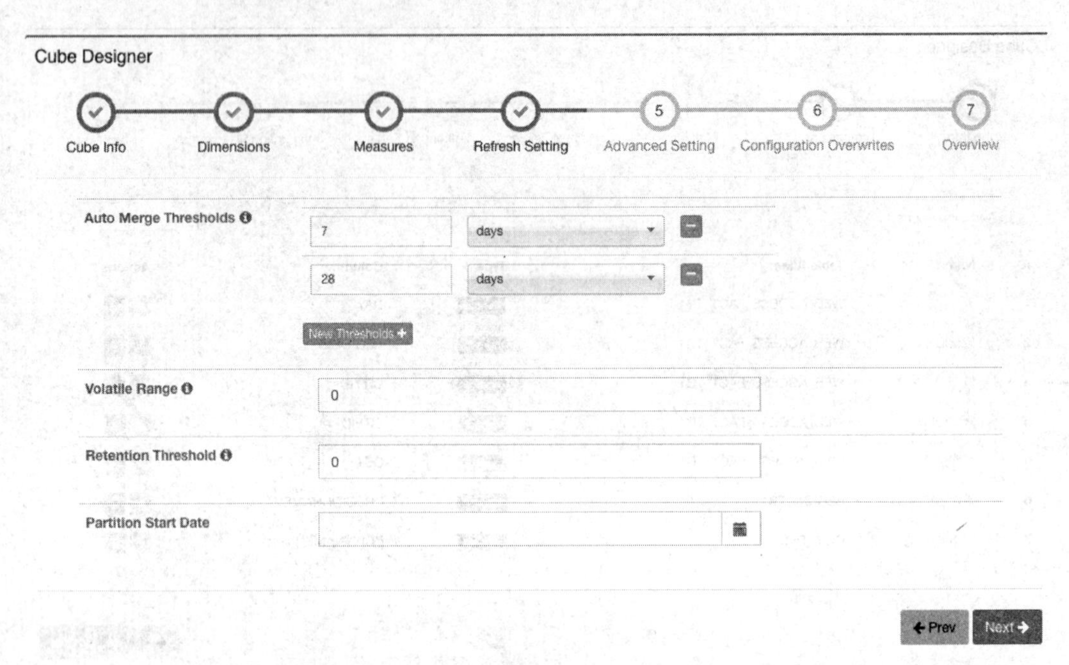

图 6-28 设计增量 Cube

设计完成后,单击"Next",进入 Kylin 高级配置界面,如图 6-29 所示。

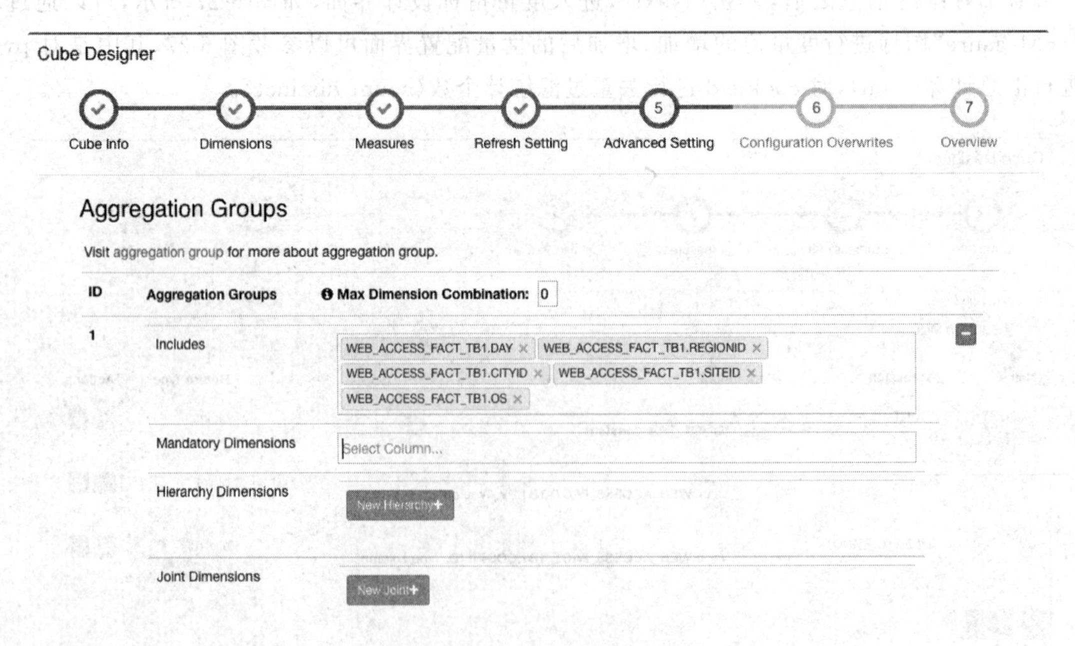

图 6-29 高级配置

图 6-29 中,Aggregation Groups(聚合组)是一种将维度进行分组,以降低维度组合数目的

手段。不同分组的维度之间组成的 Cuboid 数量会大大降低，维度组合从 2^{k+m+n} 至多能降低到 $2^k+2^m+2^n$。Group 的优化措施与查询 SQL 紧密相关，可以说是为查询定制的优化。如果查询的维度是跨组合的，那么 Kylin 需要以较大的代价从 N-cuboid 中聚合得到所需要的查询结果，这要求用户在构建 Cube 时仔细考虑和评估。

换个角度来说，维度组的设置主要是为了让不出现在一个查询中的两个维度不计算 Cuboid(通过划分到两个不同的维度组中)，这其实相当于把一个 Cube 的树结构划分成多个不同的树，可以在不降低查询性能的情况下减少 Cuboid 的计算量。

其他高级设置的内容如图 6-30 所示，此处对 HBase 的 Rowkeys(行键)进行设计，默认行键由维度的值进行组合，用户也可以继续添加行键组合的字段。一般不需要修改，使用默认值即可。

Rowkeys

Important: Dimension's positioin on HBase rowkey is critical for performance. You can drag and drop to adjust the sequence. In short, put filtering dimension before non-filtering dimension, and put high cardinality dimension before low cardinality dimension.

ID	Column	Encoding	Length	Shard By
1	WEB_ACCESS_FACT_TB1.DAY	date	0	false by default
2	WEB_ACCESS_FACT_TB1.REGIONID	dict	Col...	false by default
3	WEB_ACCESS_FACT_TB1.CITYID	dict	Col...	false by default
4	WEB_ACCESS_FACT_TB1.SITEID	dict	Col...	false by default
5	WEB_ACCESS_FACT_TB1.OS	dict	Col...	false by default

图 6-30　设计行键

单击“Next”，进行 Cube 级别的参数设置，如图 6-31 所示，单击“＋Property”可以设置 Cube 级别的参数值，此处的配置将覆盖 kylin. preperties 文件中的值，本实验暂时不需要设置，直接单击“Next”，进入最后一步。

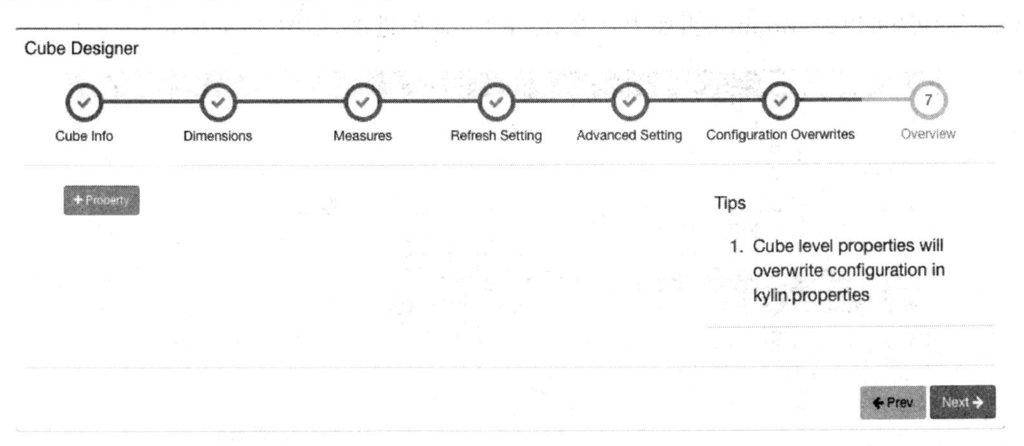

图 6-31　设置 Cube 级别

如图 6-32 所示，列出创建 Cube 的统计信息，确定没有问题后，直接单击“Save”，在弹出的界面中直接单击“Yes”,完成整个 Cube 的创建。

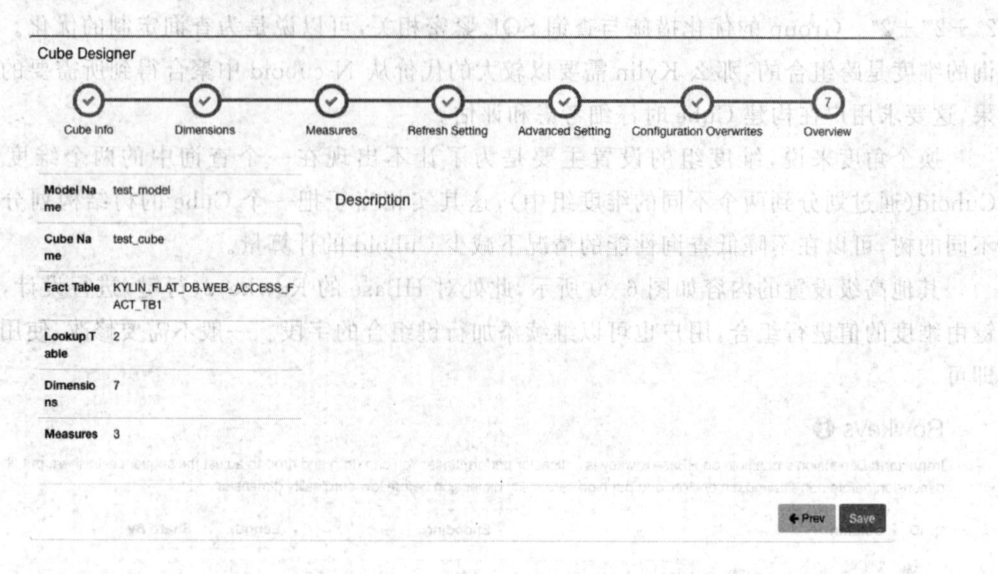

图 6-32　创建 Cube 的统计信息

上述操作完成后,查看 Cube,如图 6-33 所示,它还处于“DISABLED”状态。

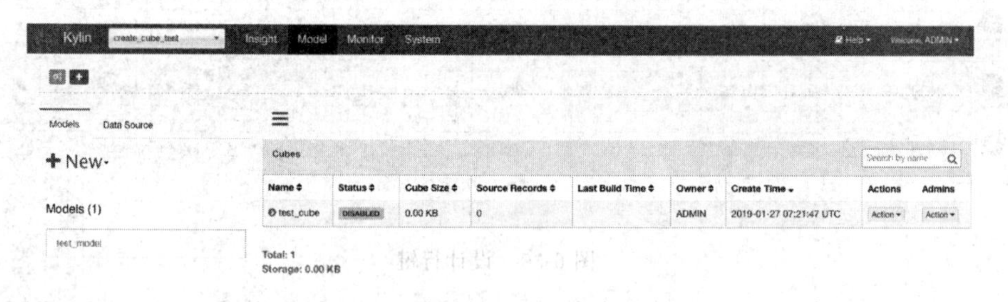

图 6-33　查看 Cube

⑤ Build Cube。Cube 创建完成后,可以开始 Build Cube,单击 Cube 的 Actions 菜单中的“Build”即可,可以在 Monitor 页面中查看 Build Cube 的进度,如图 6-34 所示。

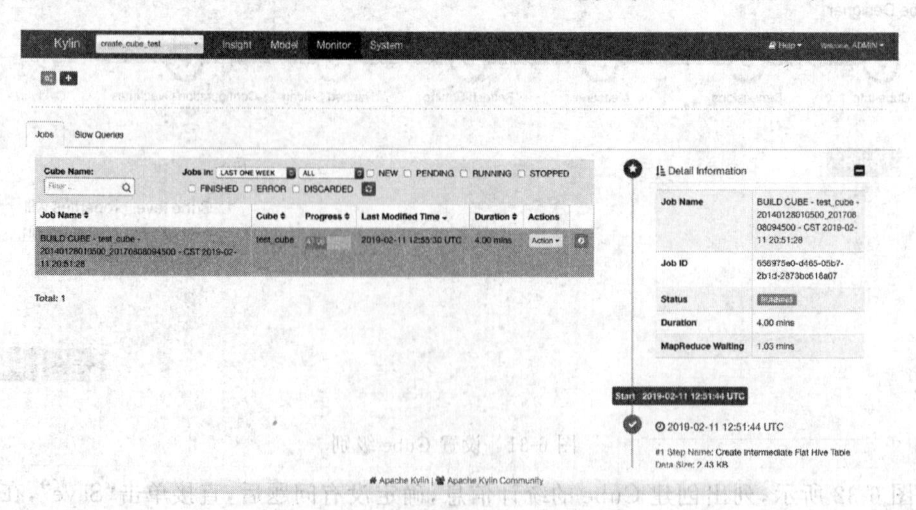

图 6-34　查看 Build Cube 的进度

Build 任务完成后，在 Model 中可以看到 Cube 已经处于“READY”状态，如图 6-35 所示。至此，本实验已经完成了 Cube 的创建，可以对其进行查询。

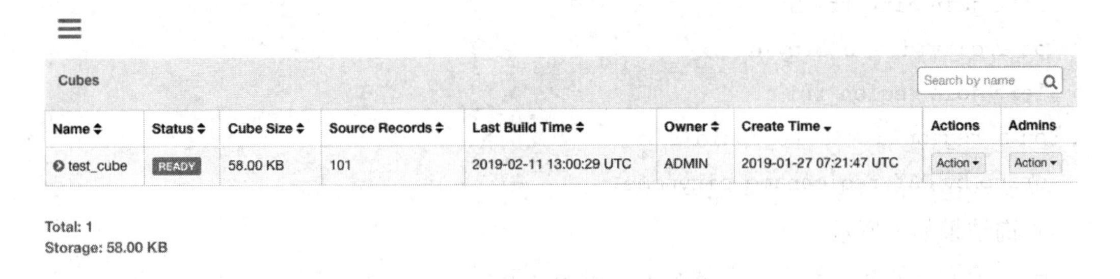

图 6-35 Cube 处于“READY”状态

⑥ 查询 Cube。通过 Insight 页面验证 Cube 构建的结果是否正常，如执行以下 SQL 语句。

```
select "DAY",regionname,cityname,sum(pv),count(distinct cookieid)
from WEB_ACCESS_FACT_TB1 a
left join CITY_TB1 b
on a.cityid = b.cityid
left join REGION_TB1 c
on c.regionid = a.regionid
group by "DAY",regionname,cityname;
```

查询结果如图 6-36 所示。

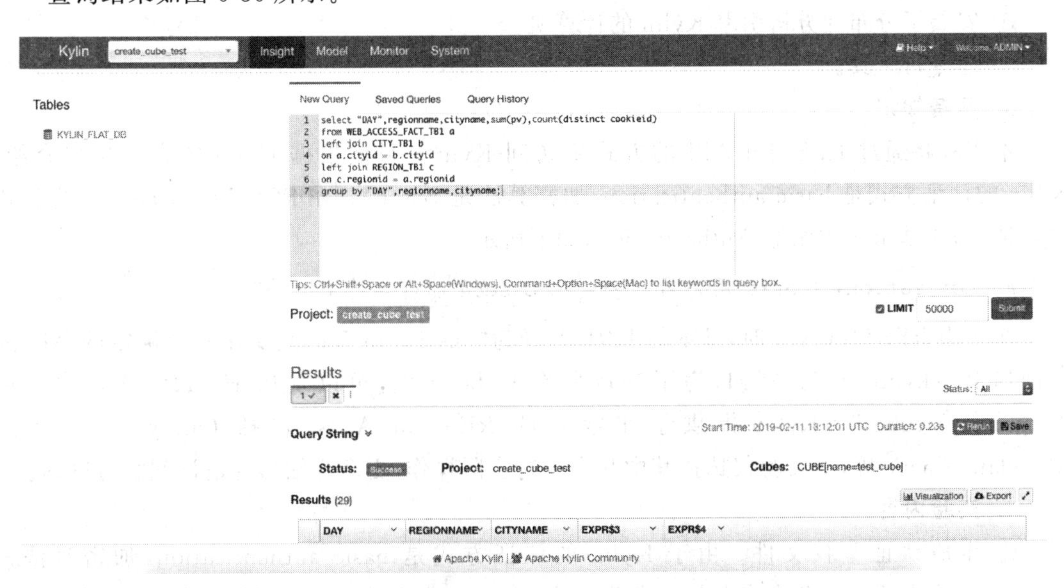

图 6-36 Cube 的查询结果

然后，在 Hive 内进行查询，SQL 语句如下所示。

```
hive > select DAY,regionname,cityname,sum(pv),count(distinct cookieid)
from web_access_fact_tb1 a
left join city_tb1 b
on a.cityid = b.cityid
left join region_tb1 c
on c.regionid = a.regionid
group by DAY,regionname,cityname;
```

查询结果如下所示。

```
Time taken:6.677 seconds, Fetched: 29 row(s)
```

可以看到,在 Kylin 内查询时间只有 0.23 s,在 Hive 中查询时间为 6.677 s,说明 Kylin 预处理后查询速度比较快。到此为止,本实验完整地介绍了整个 Cube 的创建、构建、查询,并且对每个细节都进行了较为详细的解释。

6.4 通过 RESTful 访问 Kylin

1. 实验目的

① 学会通过 RESTful 访问 Kylin。

② 了解 Kylin 的基本 RESTful API 接口以及 SQL 语句。

2. 实验要求

① 安装了分布式分析引擎 Kylin 的计算机。

② 命令行工具。

3. 预备知识

本实验将通过 RESTful API 的方式来访问 Kylin。目前根据 Kylin 的官方文档介绍,Kylin 的认证方式是 basic authentication,加密算法是 Base64。在 POST 的 header 中进行用户认证,即在请求头中添加 Authorization,如下所示。

```
Authorization:"Basic 用户名和密码的 Base64 加密字符串"
```

在使用 RESTful API 时,基本的 URL 路径为"/kylin/api",因此要在一个确定的 API 路径前加上"/kylin/api"。例如,为了获取所有 Cube 实例,可以发送 HTTP GET 请求到"/kylin/api/cubes"。Kylin 提供了非常多的 RESTful APIs,包括 Query、Cube、Job、Metadata、Cache 等,本实验仅选择其中几个进行实际操作,更多详细信息请访问官方网站。

4. 实验内容

① 生成认证鉴权文件。由于 Kylin 的认证方式是 basic authentication,加密算法是 Base64,为了方便后续操作,首先生成鉴权文件(cookiefile.txt)。需要说明的是,后续所有 curl 命令的最后都会使用"python - m json.tool"对 JSON 的输出进行格式化操作,以方便阅读。

在 Kylin 用户下执行命令,生成鉴权文件(代码中的"\"为换行符,所有行为一条指令,换行是为了方便阅读)。

```
curl -c cookiefile.txt -X POST \
-H "Authorization: Basic QURNSU46S1lMSU4 = " \
-H 'Content-Type: application/json' \
http://10.108.208.191:7070/kylin/api/user/authentication
```

返回结果如下所示。


```
{
"userDetails": {
        "password": null,
        "username": "ADMIN",
        "authorities": [
            {
                "authority": "ROLE_ADMIN"
            },
            {
                "authority": "ROLE_ANALYST"
            },
            {
                "authority": "ROLE_MODELER"
            }
        ],
        "accountNonExpired": true,
        "accountNonLocked": true,
        "credentialsNonExpired": true,
        "enabled": true
}
}
```


注意：

- Kylin 的默认用户名和密码为“ADMIN”和“KYLIN”，使用 Base64 编码后结果为“QURNSU46S1Lmsu4=”；
- -c 后的内容表示 cookie 写入的文件；
- -H 后的内容表示自定义 header 传递给服务器；
- -X 后的内容表示指定使用的请求命令。

② 查看生成的 cookiefile. txt 文件的内容。cookiefile. txt 文件的内容如图 6-37 所示，可以看到 cookiefile. txt 文件中包含 JSESSIONID。

```
# Netscape HTTP Cookie File
# https://curl.haxx.se/docs/http-cookies.html
# This file was generated by libcurl! Edit at your own risk.

#HttpOnly_10.108.208.191    FALSE    /kylin    FALSE    0    JSESSIONID
D2D3CAEF2D1F2DC401D255D1BE88D16B
```

图 6-37 cookiefile. txt 文件的内容

③ 查询 Cube 信息。在认证完成之后,可以复用 cookie 文件(不需要重新认证),向 Kylin 发送 GET 或 POST 请求。例如,查询 Cube 的信息:

```
curl -b cookiefile.txt \
-H 'Content-Type: application/json' \
http://10.108.208.191:7070/kylin/api/cubes/test_cube \
| python -m json.tool
```

返回结果如下所示。

```
{
    "uuid": "e2d91d2e-41b5-881d-a393-605e44e08a8c",
    "last_modified": 1549890029588,
    "version": "2.5.1.20500",
    "name": "test_cube",
    "owner": "ADMIN",
    "descriptor": "test_cube",
    "display_name": "test_cube",
    "cost": 50,
    "status": "READY",
    "segments": [
        {
            "uuid": "7e03a58f-d1ed-afb7-6470-a821cb32c113",
            "name": "20140128010500_20170808094500",
    …… //此处省略部分结果
    "create_time_utc": 1548573707258,
    "cuboid_bytes": null,
    "cuboid_bytes_recommend": null,
    "cuboid_last_optimized": 0,
    "snapshots": {}
}
```

可以看到查询出了 test_cube 的详细信息,包括 Cube 的 uuid、输入数据量、大小、segments 等内容,并且返回内容为 JSON 格式。

④ 通过 RESTful 查询 SQL。若要向 Kylin 发送 SQL 查询,则 POST 请求中的数据应遵从 JSON 规范,可以简单地将 JSON 格式的查询请求放到命令行中,如下所示。

```
curl -b cookiefile.txt \
-H 'Content-Type: application/json' \
-d '{"sql": "select part_dt, sum(price) as total_selled, count(distinct seller_
    id) as sellers from kylin_sales group by part_dt","offset": 0,"limit": 2,
    "acceptPartial": false,"project": "learn_kylin"}' http://10.108.208.191:
    7070/kylin/api/query \
| python -m json.tool
```

返回结果如下所示。

```
{
    "columnMetas"：[
        {
            "isNullable"：1,
            "displaySize"：0,
            "label"："PART_DT",
            "name"："PART_DT",
    ……//此处省略部分结果
    "cube"："CUBE[name = kylin_sales_cube]",
    "affectedRowCount"：0,
    "isException"：false,
    "exceptionMessage"：null,
    "duration"：142,
    "totalScanCount"：4,
    "totalScanBytes"：834,
    "hitExceptionCache"：false,
    "storageCacheUsed"：false,
    "traceUrl"：null,
    "partial"：false,
    "pushDown"：false
    "results"：[
        [
            "2012-01-01",
            "466.9037",
            "12"
        ],
        [
            "2012-01-02",
            "970.2347",
            "17"
        ]
    ],
}
```

上述结果中最后输出的“results”内容为查询结果，由于设置“limit”为2，因此输出两条记录。

⑤ 其他RESTful API请求。此处介绍加载Hive表、构建Cube和查看作业状态的RESTful API，由于该功能与WebGUI完全重合，且在WebGUI上可以更方便地实现，因此只简单地介绍代码，感兴趣的读者可以自行实践。

a. 加载 Hive 表。请求方式：POST。访问路径：http://host: port/kylin/api/tables/{table}/{project}。

```
curl -b cookiefile.txt  - X POST\
-H 'Content-Type: application/json' \
-d '{}' \
http://10.108.208.191:7070/kylin/api/tables/kylin_flat_db.region_tbl/create_cube_test \
| python -m json.tool
```

b. 构建 Cube。请求方式:PUT。访问路径:http://host:port/kylin/api/cubes/{cubeName}/rebuild。请求主体包括 startTime、endTime 和 buildType，startTime 和 endTime 以时间戳的形式显示(从 1970-01-01 00:00:00 UTC 经过的毫秒数)，buildType 可选的类型有 BUILD、MERGE 和 REFRESH。

```
curl -b cookiefile.txt  - X PUT\
-H 'Content-Type: application/json' \
-d '{
    "startTime": '1328097600000',
    "endTime": '1328184000000',
    "buildType": "BUILD",
}' \
http://10.108.208.191:7070/kylin/api/tables/kylin_flat_db.region_tbl/create_cube_test \
| python -m json.tool
```

c. 查看作业状态。请求方式:GET。访问路径:http://host:port/kylin/api/jobs/{Job ID}。

```
curl -b cookiefile.txt  - X GET\
-H 'Content-Type: application/json' \
http://10.108.208.191:7070/kylin/api/jobs/d2caccf6-903a-41a6-968a-7ce147988d7c\
| python -m json.tool
```

⑥ 使用 Python 的 requests 模块发送 HTTP 请求访问 Kylin。Python 的 requests 模块已封装 HTTP 请求与响应，Session 对象解决了认证、cookie 持久化的问题，因此适合使用场景。以下简单地演示一下发送请求。

```
$ python
>>> import requests
>>> importjson
>>> s = requests.session()
>>> headers = {'Authorization':'Basic QURNSU46S1lMSU4 ='}
>>> url = 'http://10.108.208.191:7070/kylin/api/user/authentication'
>>> s.post(url, headers = headers)
<Response [200]>
```

返回“<Response [200]>”说明请求已成功，请求所希望的响应头或数据体将随此响应返回。

接下来查询一个 Cube 的信息，如下所示。

```
>>> url2 = 'http://10.108.208.191:7070/kylin/api/cubes/kylin_sales_cube'
>>> r = s.get(url2)
>>> print(json.dumps(r.json(),sort_keys = True,indent = 4))
{
    "cost": 50,
    "create_time_utc": 0,
    "cuboid_bytes": null,
    "cuboid_bytes_recommend": null,
    "cuboid_last_optimized": 0,
    "descriptor": "kylin_sales_cube",
    "display_name": "kylin_sales_cube",
    "last_modified": 1548215653682,
    "name": "kylin_sales_cube",
    "owner": "ADMIN",
    …… //省略部分结果
    "snapshots": {},
    "status": "READY",
    "uuid": "2fbca32a-a33e-4b69-83dd-0bb8b1f8c53b",
    "version": "2.5.1.20500"
}
```

再来执行一个 SQL 查询，这里使用函数实现。

```
import requests
import json

#生成鉴权函数
def authenticate():
    url = 'http://10.108.208.191:7070/kylin/api/user/authentication'
    headers = {'Authorization': 'Basic QURNSU46S1lMSU4 ='}
    s = requests.session()
    s.headers.update({'Content-Type': 'application/json'})
    s.post(url, headers = headers)
    return s

#查询使用函数
def query(sql_str, session):
    url = 'http://10.108.208.191:7070/kylin/api/query'
```

```
    json_str = '{"sql": "'+ sql_str +'", "offset": 0, "limit": 50000,
               "acceptPartical": false, "project": "learn_kylin"}'
    r = session.post(url, data = json_str)
    results = json.dumps(r.json()['results'], sort_keys = True, indent = 4)
    return results

session = authenticate()
sql_str = 'select part_dt, sum(price) as total_selled, count(distinct seller_id)
        as sellers from kylin_sales group by part_dt'
print(query(sql_str, session))
```

返回结果如下所示。

```
[
    [
        "2012-01-21",
        "651.7491",
        "9"
    ],
    [
        "2012-01-22",
        "741.5342",
        "13"
    ],
…… //省略部分结果
]
```

实际项目中是将上述代码封装到 Python 脚本中定时调度的。

第7章

大数据可视化实验教程

7.1 ECharts数据可视化

1. 实验目的

学会在计算机上运用ECharts绘制多种可视化图表。

2. 实验要求

① ECharts。

② Linux、MacOSX或Windows系统。

③ JS编译器，推荐使用VsCode。

3. 预备知识

ECharts(Enterprise Charts)，商业级数据图表，一个纯JavaScript的图表库，可以流畅地运行在PC和移动设备上，兼容当前绝大部分浏览器(IE 6/7/8/9/10/11、Chrome、Firefox、Safari等)，底层依赖轻量级的Canvas类库ZRender，提供直观、生动、可交互、可高度个性化定制的数据可视化图表。创新的拖拽重计算、数据视图、值域漫游等特性大大增强了用户体验，赋予了用户对数据进行挖掘、整合的能力。[40]

ECharts支持折线图(区域图)、柱状图(条状图)、散点图(气泡图)、K线图、饼图(环形图)、雷达图(填充雷达图)、和弦图、力导向布局图、地图、仪表盘、漏斗图、事件河流图12类图表，同时提供标题、详情气泡、图例、值域、数据区域、时间轴、工具箱7个可交互组件，支持多图表、多组件的联动和混搭展现。

4. 实验内容

① 在官网下载合适的ECharts版本，或通过npm获取ECharts："npm install echarts --save"，或通过CDN直接引入国内的ECharts最新版本。

② 引入ECharts时，只需要同普通的JavaScript库一样用script标签引入。

```
< script src = "echarts.min.js"></script >
```

③ 绘制一个简单的图表，在绘图前需要为ECharts准备一个具备大小(宽高)的DOM容器。

```
<body>
        <!--为 ECharts 准备一个具备大小(宽高)的 DOM -->
        <div style="width:600px;height:400px;" id="chartmain"></div>
</body>
```

④ 最后,可以通过 echarts.init 方法初始化一个 ECharts 实例并通过 setOption 方法生成一个简单的柱状图,以下是完整代码。

```
<!DOCTYPE html>
<html>
<head>
<meta charset="utf-8">
<title>ECharts</title>
<!-- 引入 echarts.js -->
<script src="echarts.min.js"></script>
</head>
<body>
<!-- 为 ECharts 准备一个具备大小(宽高)的 DOM -->
<div id="main" style="width: 600px;height:400px;"></div>
<script type="text/javascript">
// 基于准备好的 DOM,初始化 ECharts 实例
var myChart = echarts.init(document.getElementById('main'));

// 指定图表的配置项和数据
var option = {
title: {
text: 'ECharts 入门示例'
},
tooltip: {},
legend: {
data:['销量']
},
xAxis: {
data: ["衬衫","羊毛衫","雪纺衫","裤子","高跟鞋","袜子"]
},
yAxis: {},
series: [{
name: '销量',
type: 'bar',
data: [5, 20, 36, 10, 10, 20]
}]
```

```
};

// 使用刚指定的配置项和数据显示图表
myChart.setOption(option);
</script>
</body>
</html>
```

最后绘制的效果图如图7-1所示。

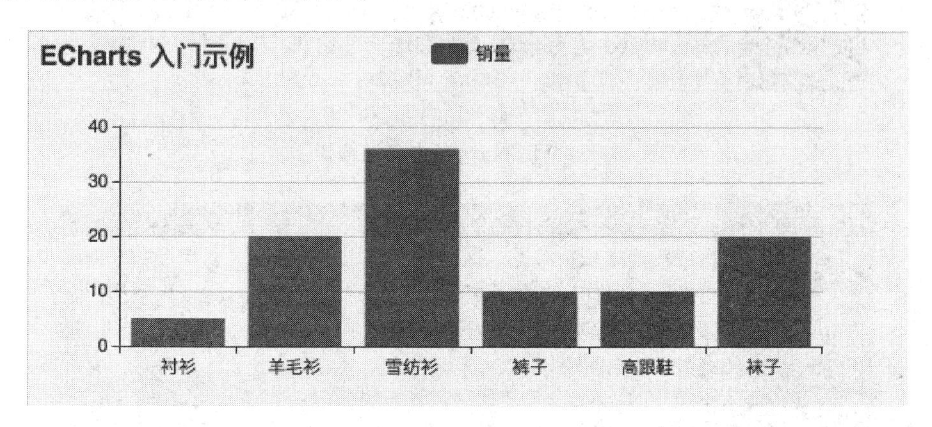

图 7-1　ECharts 绘制图展示

注意：首先，只有为ECharts准备一个具备大小的DOM容器才能将图表绘制出来；其次，图表的option选项代表不同的含义，根据配置的不同，选项的大小以及主题可以改变图表的样式。

拓展阅读

ECharts绘制关系图（知识图谱）

5. 实验作业

① 下载ECharts到本地计算机并在编辑器内绘制出效果图。

② 熟悉ECharts的整体流程以及适用的业务场景，了解每个图表的option含义，包括x轴、y轴、图例等核心组件以及数据转换成图表如何进行交互等。

6. 参考答案

实验作业②的答案如下所述。

（1）ECharts适用的业务场景

① 基于业务系统或大数据系统完成数据处理/分析后的结果数据展现。

② 在Web页面嵌入HTML的JS中应用。

③ 拥有丰富的图例和在线示例教程。

④ 同类的D3.js等有相应功能，在特殊的可视化需求中，还可以进一步考虑3D呈现的three.js、地图数据呈现的Datamaps.js等。[41]

（2）ECharts的核心组件

① ECharts基本名词解释如图7-2所示。

② ECharts的图例如图7-3所示。

名词	描述
chart	是指一个完整的图表，如折线图，饼图等“基本”图表类型或由基本图表组合而成的“混搭”图表，可能包括坐标轴、图例等
axis	直角坐标系中的一个坐标轴，坐标轴可分为类目轴和数值轴
xAxis	直角坐标系中的横轴，通常并默认为类目轴
yAxis	直角坐标系中的纵轴，通常并默认为数值轴
grid	直角坐标系中除坐标轴外的绘图网格
legend	图例
dataRange	值域选择，常用于展现地域数据时选择值域范围
toolbox	辅助工具箱
tooltip	气泡提示框，用于展现更详细的数据
dataZoom	数据区域缩放，常用于展现大数据时选择可视范围
series	数据系列

图 7-2　ECharts 基本名词解释

名词	描述
line	折线图，堆积折线图，区域图，堆积区域图
bar	柱形图（纵向），堆积柱形图，条形图（横向），堆积条形图
scatter	散点图，气泡图，大规模散点图
k	K线图，蜡烛图
pie	饼图，圆环图
radar	雷达图，填充雷达图
chord	和弦图
force	力导布局图

图 7-3　ECharts 图例类型

(3) ECharts 的数据交互

交互是从数据中发掘信息的重要手段，总览为先、缩放过滤、按需查看细节是数据可视化交互的基本需求。

ECharts 一直在“交互”的路上前进，legend、visualMap、dataZoom、tooltip 等组件以及图表附带的漫游、选取等操作提供了数据筛取、视图缩放、展示细节等功能。

ECharts 3 对这些组件进行了广泛增强，如支持在数据的各种坐标轴、维度进行数据过滤、缩放，以及在更多的图中采用这些组件。

ECharts 3 开始加强了对多维数据的支持，除加入了平行坐标等常见的多维数据可视化工具外，对于传统的散点图等，传入的数据也可以是多个维度的，配合视觉映射组件 visualMap 提供的丰富的视觉编码，能够将不同维度的数据映射到颜色、大小、透明度、明暗度等不同的视觉通道。

7.2　Plotly 数据可视化

1. 实验目的

学会运用 Plotly 绘制多种可视化图表。

2. 实验要求

① Plotly。

② Linux、MacOSX 或 Windows 系统。

③ Jupyter Notebook 开发工具。

3. 预备知识

Plotly 是一个非常著名且强大的开源数据可视化框架，它通过构建基于浏览器显示的 Web 形式的可交互图表来展示信息，可创建数十种精美的图表和地图，可以供 JS、Python、R、DB 等使用。[42]

Plotly 绘图底层使用的是 plotly. js，plotly. js 基于 D3. js、stack. gl（WebGL 组件库，由 Plotly 团队的 Mikola Lysenko 带领开发）和 SVG，用 JavaScript 在网页上实现类似于 MATLAB 和 Python Matplotlib 的图形展示功能，支持数十种图形，包括 2D 和 3D 图形，交互流畅，可以满足一般科学计算的需要。Plotly 支持 20 种基本图表、12 种统计和海运方式图、2 种财务图表、21 种科学图表等，还支持地图、报告生成、流动图表以及连接数据库。

4. 实验内容

拓展阅读

Plotly 在线绘图

① Plotly 中绘制图像有在线和离线两种方式，本实验仅介绍离线绘图的方式。离线绘图包括 plotly. offline. plot()和 plotly. offline. iplot()两种方法，前者是以离线的方式在当前工作目录下生成 HTML 格式的图像文件，并自动打开；后者是在 Jupyter Notebook 中专用的方法，即将生成的图形嵌入 ipynb 文件中，本实验采用后一种方式（注意，在 Jupyter Notebook 中使用 plotly. offline. iplot()之前，需要运行 plotly. offline. init_notebook_mode()以完成绘图代码的初始化，否则会报错）。

② graph 对象：Plotly 中的 graph_objs 是 Plotly 下的子模块，用于导入 Plotly 中的所有图形对象，在导入相应的图形对象之后，便可以根据需要呈现的数据和自定义的图形规格参数来定义一个 graph 对象，再输入 plotly. offline. iplot()中进行最终的呈现。

```
import plotly
import plotly.graph_objs as go
```

③ 构造 traces：在根据绘图需求从 graph_objs 中导入相应的 obj 之后，接下来需要基于待展示的数据，为指定的 obj 配置相关参数，这在 Plotly 中称作构造 traces，一张图中可以叠加多个 trace。

```
##创建仿真数据
N = 100
random_x = np.linspace(0, 1, N)
random_y0 = np.random.randn(N) + 5
random_y1 = np.random.randn(N)
random_y2 = np.random.randn(N) - 5
##构造 trace0
trace0 = go.Scatter(
    x = random_x,
    y = random_y0,
    mode = 'markers',
```

```
    name = 'markers'
)
##构造 trace1
trace1 = go.Scatter(
    x = random_x,
    y = random_y1,
    mode = 'lines+markers',
    name = 'lines+markers'
)
##将所有 trace 保存在列表中
data = [trace0, trace1]
```

④ 定义 Layout:Plotly 中图像的图层元素与底层的背景、坐标轴等是相互独立的,在定义好绘制图像需要的对象之后,可以直接绘制,但如果想要在背景图层上有更多自定义化的内容,就需要定义 Layout()对象。

```
##创建 Layout 对象,对横、纵坐标轴的标题进行一定的设置
layout = go.Layout(xaxis={
    'title':'这是横坐标轴',
    'titlefont':{
        'size':30
    }
},yaxis={
    'title':'这是纵坐标轴',
    'titlefont':{
        'size':40
    }
})
```

⑤ 启动绘图。

```
plotly.offline.init_notebook_mode()
plotly.offline.iplot(data, filename='scatter-mode')
```

⑥ 最后绘制的效果图如图 7-4 所示。

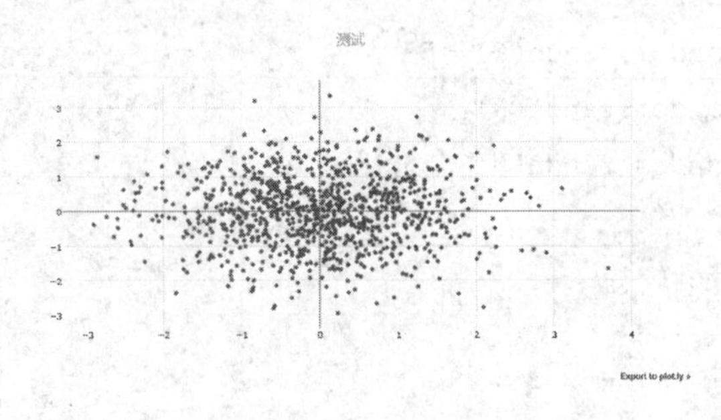

图 7-4　绘制效果图

⑦ 完整代码如下所示。

```
import plotly
import plotly.graph_objs as go
import numpy as np
'''构造1000个服从二维正态分布的模拟数据'''
N = 1000
random_x = np.random.randn(N)
random_y = np.random.randn(N)
'''构造trace,配置相关参数'''
trace = go.Scatter(
    x = random_x,
    y = random_y,
    mode = 'markers'
)
'''将trace保存于列表之中'''
data = [trace]
'''创建Layout对象'''
layout = go.Layout(title='测试',
                   titlefont={
                      'size':20,
                      'color':'9ed900'#将标题字体颜色设置为葱绿色
                   }
)

'''将graph部分和layout部分组合成figure对象'''
fig = go.Figure(data=data, layout=layout)
'''启动绘图直接绘制figure对象'''
plotly.offline.init_notebook_mode()
plotly.offline.iplot(fig,filename='basic-scatter')
```

5. 实验作业

① 运用Plotly在开发工具内绘制出效果图。

② 熟悉Plotly的整体流程以及适用的业务场景,了解Layout的自定义内容。

6. 参考答案

实验作业②的答案如下所述。

(1) 文字

文字是一幅图中十分重要的组成部分,Plotly凭借其强大的绘图机制对一幅图中的文字进行了细致的划分,可以非常有针对性地对某一个组件部分的字体进行个性化的设置。

① 全局文字。

- font:字典型,用于控制图像中全局字体的部分。

- family：str 型，用于控制字体，默认为 Open Sans，可选项有 Verdana、Arial、Sans-serif 等，具体可参考官网的说明文档。
- size：int 型，用于控制字体大小，默认为 12。
- color：str 型，传入十六进制色彩，默认为＃444。

② 标题文字。

- title：str 型，用于控制图像的主标题。
- titlefont：字典型，用于独立控制标题字体的部分。
- family：同全局文字中的 family，用于单独控制标题字体。
- size：int 型，控制标题的字体大小。
- color：同全局文字中的 color。

(2) 坐标轴

- xaxis(yaxis)：字典型，控制横坐标(纵坐标)的各属性，如 color 为 str 型，传入十六进制色彩，控制横坐标上所有元素的基础颜色。
- title：str 型，设置坐标轴上的标题。
- type：str 型，用于控制横坐标轴类型。“-”表示根据输入数据自适应调整，“linear”表示线性坐标轴，“log”表示对数坐标轴，“date”表示日期型坐标轴，“category”表示分类型坐标轴，默认为“-”。

(3) 图例

- showlegend：bool 型，控制是否绘制图例。
- legend：字典型，用于控制与图例相关的所有属性，包括图例背景颜色、图例边框颜色、图例文字部分的字体等。

(4) 其他

- width：int 型，控制图像的像素宽度，默认为 700。
- height：int 型，控制图像的像素高度，默认为 450。
- margin：字典型，控制图像边界的宽度等。

7.3 D3.js 绘制知识图谱

1. 实验目的

学会使用 D3.js 绘制知识图谱。

2. 实验要求

① D3.js。

② Linux、MacOSX 或 Windows 系统。

③ JS 编译器，推荐使用 VsCode。

3. 预备知识

D3 的全称是 Data-Driven Documents，直译为“数据驱动的文档”。D3.js 是一个基于数据来操作文档的 JavaScript 库。D3 允许用户绑定任意数据到 DOM，然后根据数据来操作 DOM，创建交互式的图表。[41]

4. 实验内容

(1) D3.js 的引入

D3 类库的文件名为 D3.js,只有一个文件,所有的对象、函数、变量都写在此文件中,只要在<script>中引入此文件即可,有以下两种引入方式。

① 将库文件下载下来,放在工程目录中引用。D3 的官方网站为 http://d3js.org/。[41] 找到下载链接,文件名为 D3.zip,解压缩后得到 3 个文件:

- d3.js,开发项目时为了调试方便,输出更多信息,可使用此文件;
- d3.min.js,是 d3.js 压缩后的版本,去掉了空格等,功能完全一样,但体积较小,浏览器读取的速度会快不少,适合发布时使用;
- LICENSE,是版权许可证文件。

在 HTML 文件中引入 d3.js,代码如下所示。

```
<script src = "d3/d3.js" charset = "utf - 8"></script>
```

属性"charset"指定解析文件的编码方式。特别注意,在指定目录时,地址"d3/d3.js"之前没有斜杠"/"。如果加了斜杠,就表示在服务器根目录下的 d3 文件夹中的文件 d3.js。不加斜杠,表示当前 HTML 文件所在目录下。如果是上一级目录,路径为"../d3/d3.js"。

② 通过网络引入在线文件。这种方法无须下载文件,但只有保持网络连接才可使用。

```
<script src = "https://d3js.org/d3.v3.min.js"></script>
```

(2)绘制矢量图

一般来说,D3 是在 SVG 画板上作图的,代码如下所示。

```
let width = 800;
let height = 800;
let svg = d3.select('body')          //选择 body
.append('svg')                       //在 body 中添加<svg>
.attr('width',width)                 //设置<svg>的宽度属性
.attr('height',height);              //设置<svg>的高度属性
```

函数 select()用于选择对象,append()用于添加元素,attr()用于给指定属性赋值。

(3) 设置数据,包括节点和边

具体代码如下所示。

```
let data = {
            'nodes':[
                      {'food':'西红柿'},
                      {'food':'茄子'},
                      {'food':'胡萝卜'},
                      {'food':'辣椒'},
                      {'food':'土豆'},
                      {'food':'白菜'},
                      {'food':'粉条'},
                      {'food':'山药'}
            ],
```

```
          'links':[
                  {source:0,target:1},
                  {source:0,target:2},
                  {source:0,target:3},
                  {source:0,target:4},
                  {source:0,target:5},
                  {source:0,target:6},
                  {source:0,target:7}
          ]
   };
let nodes = data.nodes;
let links = data.links;
```

(4) 布局(数据转换)

定义一个力导向图的布局，代码如下所示。

```
let force = d3.layout.force()
    .nodes(nodes)                    //设定节点数组
    .links(links)                    //设定关系数组
    .size([width,height])            //设定作用域的范围
    .linkDistance(300)               //设定关系连线的长度
    .distance(280)
    .charge(-1500);                  //节点间的相互作用力
```

然后，使力学作用生效：

```
force.start(); //开始引力作用
```

(5) 绘制

定义一个颜色生成器，由 D3 提供：

```
let color = d3.scale.category20(); //20 表示可以生成 20 种不同的颜色
```

绘制节点间的关系连线：

```
let linksLine = svg.selectAll('line')
                   .data(links)
                   .enter()
                   .append('line')
                   .style('stroke','#ccc')
                   .style('stroke-width',2);
```

绘制节点：

```
let nodesCic = svg.selectAll('circle')
                  .data(nodes)
                  .enter()
```

```
                .append('circle')
                .attr('r',25)
                .style('fill',function(d,i){
                //根据自身的索引号,自动随机获取一种背景颜色
                    return color(i);
                })
                .call(force.drag); //调用 drag 函数,使得节点可以拖动
```

给节点添加名称:

```
let nodesTitle = svg.selectAll('text')
                .data(nodes)
                .enter()
                .append('text')
                .style('fill','#fff')
                .attr("text-anchor", "middle") //使得文字居中显示在节点上
                .attr('dx',20)                 //自定义文字的 x 坐标
                .attr('dy',5)                  //自定义文字的 y 坐标
                .text(function(d){
                        return d.food;
                });
```

绘制整个知识图谱:

```
force.on('tick',tick);
            function tick(){
                // 更新连线的坐标
                linksLine.attr('x1',function(d){return d.source.x;})
                .attr('y1',function(d){return d.source.y;})
                .attr('x2',function(d){return d.target.x;})
                .attr('y2',function(d){return d.target.y;});
                // 更新节点坐标
                nodesCic.attr('cx',function(d){return d.x;})
                .attr('cy',function(d){return d.y});
                // 更新文字坐标
                nodesTitle.attr('x',function(d){return d.x})
                .attr('y',function(d){return d.y + 5});

            }
```

(6) 绘制效果图展示

D3.js 绘制的知识图谱如图 7-5 所示。

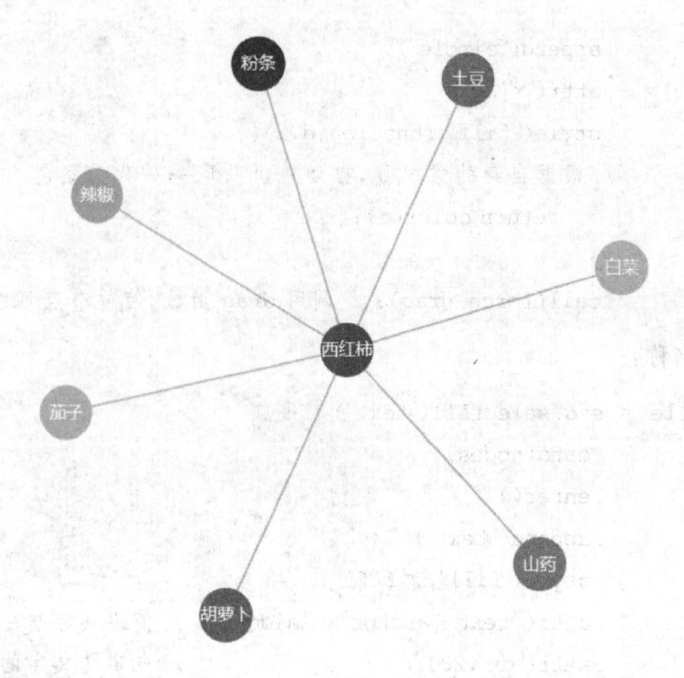

图 7-5 D3.js 绘制的知识图谱

5. 实验作业

① 根据实验内容，使用 D3.js 绘制出图 7-5 所示的知识图谱。

② 绘制更加丰富的知识图谱，懂得绘制知识图谱的真正意义，分析图谱之间的复杂关系。

6. 参考答案

① 实验内容完整代码如下所示。

```
<! DOCTYPE html>
<html lang = "en">

<head>
    <meta charset = "UTF-8">
    <script src = "https://d3js.org/d3.v3.min.js"></script>
    <script src = "d3/d3.js" charset = "utf-8"></script>
</head>

<body>

    <script>
        let data = {
            'nodes': [{
                        'food': '西红柿'
            },
```

```
        {
                'food':'茄子'
        },
        {
                'food':'胡萝卜'
        },
        {
                'food':'辣椒'
        },
        {
                'food':'土豆'
        },
        {
                'food':'白菜'
        },
        {
                'food':'粉条'
        },
        {
                'food':'山药'
        }
    ],

    'links':[{
                source:0,
                target:1
        },
        {
                source:0,
                target:2
        },
        {
                source:0,
                target:3
        },
        {
                source:0,
                target:4
        },
```

```
        {
                    source：0,
                    target：5
        },
        {
                    source：0,
                    target：6
        },
        {
                    source：0,
                    target：7
        }
    ]
};

let nodes = data.nodes;
let links = data.links;

let width = 800;
let height = 800;

let svg = d3.select('body')
          .append('svg')
          .attr('width', width)
          .attr('height', height);

let force = d3.layout.force()
            .nodes(nodes)             //设定节点数组
            .links(links)             //设定关系数组
            .size([width, height])  //设定作用域的范围
            .linkDistance(300)        //设定关系连线的长度
            .distance(280)
            .charge(-1500);           //节点间的相互作用力

force.start(); //开始引力作用

// 定义一个颜色生成器，由 D3 提供
let color = d3.scale.category20(); //20 表示可以生成 20 种不同的颜色
```

```
// 下面开始绘制
// 绘制节点间的关系连线
let linksLine = svg.selectAll('line')
                .data(links)
                .enter()
                .append('line')
                .style('stroke', '#ccc')
                .style('stroke-width', 2);

// 绘制节点
let nodesCic = svg.selectAll('circle')
               .data(nodes)
               .enter()
               .append('circle')
               .attr('r', 25)
               .style('fill', function (d, i) {
                   return color(i); //根据自身的索引号,自动随机获取一种背
                                      景颜色
               })
               .call(force.drag);   //调用 drag 函数,使得节点可以拖动

// 给节点添加名称
let nodesTitle = svg.selectAll('text')
                 .data(nodes)
                 .enter()
                 .append('text')
                 .style('fill', '#fff')
                 .attr("text-anchor", "middle") //使得文字居中显示在节点上
                 .attr('dx',20)                 //自定义文字的 x 坐标
                 .attr('dy',5)                  //自定义文字的 y 坐标
                 .text(function (d) {
                     return d.food;
                 });

force.on('tick', tick);

function tick() {
        // 更新连线的坐标
        linksLine.attr('x1', function (d) {
```

```
                return d.source.x;
            })
            .attr('y1', function (d) {
                return d.source.y;
            })
            .attr('x2', function (d) {
                return d.target.x;
            })
            .attr('y2', function (d) {
                return d.target.y;
            });

        // 更新节点坐标
        nodesCic.attr('cx', function (d) {
                return d.x;
            })
            .attr('cy', function (d) {
                return d.y
            });

        // 更新文字坐标
        nodesTitle.attr('x', function (d) {
                return d.x
            })
            .attr('y', function (d) {
                return d.y + 5
            });

    }
    </script>

  </body>

</html>
```

② 实验相关材料可以查阅 D3 官网进行系统的学习。

第8章

大数据综合实验案例

本案例涉及数据采集、预处理、存储、查询和可视化分析、数据挖掘等数据处理全流程中的各种典型操作与应用。本案例能够帮助读者综合运用大数据课程知识和各种工具软件，实现数据全流程操作。本案例中使用的数据集合源码都可以在 www. buptpress. com 下载。

8.1 案例简介

本节将介绍本案例的目的、预备知识、硬件要求、软件工具、数据集等内容。

1. 实验目的

① 熟悉 Linux、MySQL、Hadoop、HBase、Hive、Sqoop 等系统和软件的安装和使用。

② 了解大数据处理的基本流程。

③ 熟悉数据预处理方法。

④ 熟悉使用 Spark MLlib。

2. 预备知识

本实验需要实验者已经学习过大数据相关课程，了解大数据相关技术的基本概念与原理，了解大数据处理架构 Hadoop 的关键技术及其基本原理、数据仓库的概念与原理、关系型数据的概念与原理、Spark MLlib 的使用方法、Scala 语言的概念与应用等。

3. 硬件要求

本案例可以在单机上完成，也可以在集群环境下完成。在单机上完成本案例时，建议计算机硬件配置：500 GB 以上硬盘，8 GB 以上内存。

4. 软件工具

① Sqoop。

② IntelliJ IDEA。

5. 数据集

数据集包括 movie 表和 score 表，movie 表中包含 1 696 部电影的基本信息，score 表中包含超过 50 万条用户评分，8.3 节将详细介绍数据集的数据组成。

8.2 实验步骤

本案例共包含4个步骤，如下所述。

① 下载数据集并导入MySQL数据库〔也可利用本书提供的爬虫(Python)源码，自行爬取更大的数据集〕。

② 使用Sqoop将MySQL中的数据导入数据仓库。

③ Hive数据分析。

④ 利用Spark MLlib实现电影相似度计算。

8.3 数据集下载

本案例使用的数据是利用爬虫技术从豆瓣网爬取的1 696部电影和超过50万条用户评分数据，采用的爬虫框架是Scrapy，因此需要读者有一定的Python基础。本书会提供爬虫程序的源代码，当然读者也可以在本书提供的网址直接下载数据集，并将数据集导入MySQL数据库中。

1. 实验数据简介

数据集包括两个表，即movie表和score表。movie表主要包含电影名称、电影类别、电影海报等信息。score表包含三个字段，分别是电影ID、用户名以及用户对电影的评分。表8-1和表8-2给出了两个数据表的各字段含义。

表8-1 movie表各字段含义

id	电影ID
title	电影名称
loc	电影出品地
cate	电影类别
url	电影豆瓣地址
cover	电影海报
rate	电影评分

表8-2 score表各字段含义

mid	电影ID
uname	用户名
rate	评分

2. 实验数据下载

本案例采用的数据集为data.zip，其中包含两个sql文件，分别为movie.sql和score.sql。读者可以通过使用浏览器访问www.buptpress.com下载data.zip文件，然后使用scp命令将data.zip上传到Linux系统，命令如下所示。

```
$ scp D:\download\data.zp root@10.108.211.22:~/下载/data.zip
```

进入存放 data.zip 的文件夹内，使用 tar 命令将该文件解压到对应的文件夹下，命令如下所示。

```
$ cd ~/下载
$ mkdir dataset
$ sudo tar -zxvf data.zip -C ~/下载/dataset
```

可以看到在 dataset 目录下有两个文件 movie.sql 和 score.sql，movie.sql 是电影表的执行文件，score.sql 为评分表的执行文件。

3. 数据集导入 MySQL

准备好 sql 文件后，可以使用命令行将数据集导入 MySQL 数据库中。首先通过命令行登录 MySQL 数据库，执行以下命令。

```
$ mysql -u username -p password
```

注意：读者应将上述命令中的"username"替换成自己的数据库用户名，将"password"替换成自己的登录密码。

接下来需要新建一个库，执行以下命令。

```
mysql > CREATE DATABASE IF NOT EXISTS douban default charset utf8 COLLATE
    utf8_general_ci;
```

可以看到已经创建了一个名为 douban 的库，接下来需要将两个 sql 文件导入 douban 库中，命令如下所示。

```
mysql > use douban;
mysql > source ~/下载/dataset/movie.sql;
mysql > source ~/下载/dataset/score.sql;
```

经过上述操作后，在新建的数据库 douban 中已经存在 movie、score 两个表，可以通过命令"show tables;"查看库中的数据表，结果如图 8-1 所示。

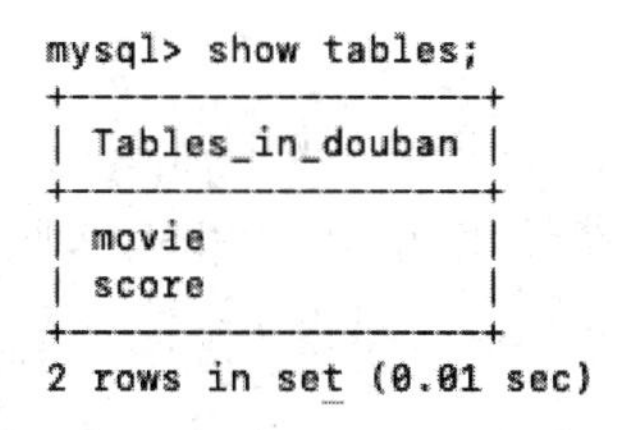

图 8-1 当前库中的表

图 8-1 中的结果说明数据已经成功导入 MySQL 数据库中，接下来对数据进行简单介绍。

8.4 数据集导入数据仓库 Hive

Sqoop 是一款开源的工具，主要用于在 Hadoop 与传统的关系数据库之间进行数据的传递，可以将一个关系数据库（如 MySQL、Oracle 等）中的数据导入 Hadoop 的 HDFS、HBase、

Hive 中,也可以将 Hadoop 的数据导入关系数据库中。Sqoop 项目开始于 2009 年,最早作为 Hadoop 的一个第三方模块存在,后来为了让使用者能够快速部署,也为了让开发人员能够更快地迭代开发,Sqoop 独立成为一个 Apache 项目。[43]

Sqoop 的安装环境如下所述。

① 操作系统:Linux 系统。

② Sqoop:1.4.7 版本。

③ Hadoop:2.7.1 版本。

④ MySQL:5.7.15 版本。

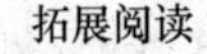

MySQL 与 Hadoop 数据同步

其中需要注意的是,Sqoop 1.x 版本和 Sqoop 2.x 版本完全不兼容,本书使用 Sqoop 1.4.7。其他软件的版本可以根据自身情况而定,不一定要和本书一致。

1. 下载安装文件

登录 Linux 系统,使用 wget 命令下载 Sqoop 安装文件 sqoop-1.4.7.bin_hadoop-2.6.0.tar.gz。执行以下命令,下载安装文件。

```
$ wget https://www-eu.apache.org/dist/sqoop/1.4.7/sqoop-1.4.7.bin_hadoop-2.6.0.tar.gz
```

进入存放 Sqoop 安装文件的文件夹,执行以下命令,完成安装文件解压缩。

```
$ cd ~/下载
$ sudo tar -zxvf sqoop-1.4.7.bin_hadoop-2.6.0.tar.gz -C /usr/local
```

上述命令将 Sqoop 安装文件解压到了/usr/local 目录下,接下来将解压的文件的名称修改为 sqoop,以简化操作,命令如下所示。

```
$ cd /usr/local
$ sudo mv sqoop-1.4.7.bin_hadoop-2.6.0 sqoop
```

2. 修改配置环境

执行以下命令,将/usr/local/sqoop/conf 目录下的配置文件 sqoop-env-template.sh 复制一份,并重命名为 sqoop-env.sh。

```
$ cd /usr/local/sqoop/conf/
$ cat sqoop-env-template.sh >> sqoop-env.sh
```

执行以下命令,使用 vim 编辑器打开 sqoop-env.sh 文件进行编辑。

```
$ cd /usr/local/sqoop/conf/
$ vim sqoop-env.sh
```

在 sqoop-env.sh 文件中添加以下配置信息。

```
export HADOOP_HOME = /usr/hdp/3.0.1.0-187/hadoop
export HBASE_HOME = /usr/hdp/current/hbase-client
export HIVE_HOME = /usr/hdp/current/hive-client
#export ZOOCFGDIR = ${ZOOCFGDIR:-/etc/zookeeper/conf
```

3. 配置环境变量

使用 vim 编辑器打开 /etc/profile 文件,命令如下所示。

```
$ vim/etc/profile
```

在该文件中追加以下内容。

```
export SQOOP_HOME = /usr/local/sqoop
export PATH = $PATH: $SBT_HOME/bin: $SQOOP_HOME/bin
export CLASSPATH = $CLASSPATH: $SQIIP_HOME/lib
```

保存该文件并退出 vim 编辑器，再执行以下命令使环境变量立即生效。

```
$ source /etc/profile
```

4. 添加 MySQL 驱动程序

首先安装 MySQL，然后下载 MySQL 驱动程序，使用 wget 命令下载 mysql-connector-java-6.0.6.jar，命令如下所示。

```
$ cd ~/下载
$ wgethttp://central.maven.org/maven2/mysql/mysql-connector-java/6.0.6/
mysql-connector-java-6.0.6.jar
```

接下来把上述文件复制到 $SQOOP_HOME/lib 目录下，命令如下所示。

```
$ cp  ~/下载/mysql-connector-java-6.0.6.jar  /usr/local/sqoop/lib
```

5. 测试与 MySQL 的连接

首先要确保 MySQL 已经开启，然后测试 Sqoop 与 MySQL 之间的连接是否成功，命令如下所示。

```
$ sqoop list-databases --connect jdbc:mysql://127.0.0.1:3306/ --username
 root -P
```

读者可根据自身情况填写 IP 地址、端口号、数据库用户名，此外，系统会提示输入密码。上述命令的执行结果如图 8-2 所示。

```
[root@slave2 ~]# sqoop list-databases --connect jdbc:mysql://127.0.0.1:3306/ --username root -P
Warning: /usr/hdp/3.0.1.0-187/accumulo does not exist! Accumulo imports will fail.
Please set $ACCUMULO_HOME to the root of your Accumulo installation.
SLF4J: Class path contains multiple SLF4J bindings.
SLF4J: Found binding in [jar:file:/usr/hdp/3.0.1.0-187/hadoop/lib/slf4j-log4j12-1.7.25.jar!/org/slf4j/impl/StaticLoggerBinder.class]
SLF4J: Found binding in [jar:file:/usr/hdp/3.0.1.0-187/hive/lib/log4j-slf4j-impl-2.10.0.jar!/org/slf4j/impl/StaticLoggerBinder.class]
SLF4J: See http://www.slf4j.org/codes.html#multiple_bindings for an explanation.
SLF4J: Actual binding is of type [org.slf4j.impl.Log4jLoggerFactory]
19/10/29 17:39:11 INFO sqoop.Sqoop: Running Sqoop version: 1.4.7.3.0.1.0-187
Enter password:
19/10/29 17:39:20 INFO manager.MySQLManager: Preparing to use a MySQL streaming resultset.
information_schema
daping
data_management
data_ware
douban
druid
hive
jishengwei
jishengwei_test
luo
mysql
performance_schema
preprocess
sentry
sys
visualization
```

图 8-2 测试 Sqoop 连接

如果能够看到 MySQL 数据库中的数据库列表展示到命令行上，就表示 Sqoop 安装成功。例如，从图 8-2 中可以看到在最后几行包含以下数据库列表。

```
information_schema
mysql
performance_schema
sys
```

6. 将 MySQL 中的数据导入 HDFS

Sqoop 安装完成后，接下来就可以将存储在 MySQL 中的数据集导入 HDFS，在命令行输入以下命令执行数据导入操作，首先导入 movie 表。

```
$ sqoop import
--connect jdbc:mysql://localhost :3306/douban\
--username root \
--password pwd \
--table movie \
--columns "id,title,loc,cate,url,cover,rate" \
--target-dir/douban/movie \
----fields-terminated-by '|'
-m 1
```

以上操作完成了 movie 表的导入，接下来导入 score 表，score 表的导入相对于 movie 表的导入会复杂一些，因为 score 表中有重复数据，实验中需要过滤重复数据，执行以下命令。

```
$ sqoop import
--connect jdbc:mysql://localhost :3306/douban\
--username root \
--password pwd \
--query 'select distinct mid,uname,rate from score where $CONDITIONS' \
--target-dir/douban/score \
----fields-terminated-by '|'
-m 1
```

读者需要注意，--connect、--username、--password 后的参数都要根据自己的数据库进行设置。

上述命令的具体含义如表 8-3 所示。

表 8-3 命令的具体含义

命令	含义
sqoop import	数据从 MySQL 导入 HDFS。相反为“sqoop export”
--connect jdbc:mysql//:localhost:3306/douban	数据库链接地址
--username root	数据库的登录用户名
--password 123456	数据库登录密码
--table movie	数据库中即将被导出的表，不能与“--query”一起使用

续表

命令	含义
--colums "id,title,loc,cate,url,cover,rate"	选择表中被导出的列
--query ""	导入的查询语句,不能与"--table"一起使用
--target-dir /douban/score	指定导入的目标 HDFS 文件夹
--delete-target-dir	如果目标文件夹已存在,则删除
--fields-terminated-by '|'	MySQL 中被导出数据的字段分割符
-m 1	使用 1 个 map 任务并行导入数据

经过上述操作,数据集已经成功导入 HDFS,可使用"hdfs dfs -ls"命令进行查看,结果如图 8-3 所示。

```
[root@slave2 ~]# hdfs dfs -ls /douban
Found 2 items
drwxr-xr-x   - hdfs hdfs          0 2019-03-16 15:40 /douban/movie
drwxr-xr-x   - hdfs hdfs          0 2019-02-27 17:09 /douban/score
```

图 8-3　查看 douban 目录下的文件列表

接下来使用"hdfs dfs -ls /douban/movie"命令查看 movie 文件下的文件,结果如图 8-4 所示。

```
[root@slave2 ~]# hdfs dfs -ls /douban/movie
Found 1 items
-rw-r--r--   3 hdfs hdfs          0 2019-02-28 15:19 /douban/movie/_SUCCESS
-rw-r--r--   3 hdfs hdfs     289450 2019-02-28 15:19 /warehouse/tablespace/external/hive/douban.db/movie/part-m-00000
```

图 8-4　查看 movie 目录下的文件

"part-m-00000"便是被导入的数据文件,使用"hdfs dfs -cat /douban/movie/part-m-00000 -10"就可以查看文本的前十行,结果如图 8-5 所示,从中可以看到,每个字段之间使用"|"作为分割符。

```
[root@slave2 ~]# hdfs dfs -cat /douban/movie/part-m-00000 | head -10
10001432|颠倒的帕特玛|日本|动画|https://movie.douban.com/subject/10001432/|https://img1.doubanio.com/view/photo/s_ratio_poster/public/p2043944718.jpg|7.9
10174444|剧场版魔法少女小圆 前篇 起始的故事|日本|动画|https://movie.douban.com/subject/10174444/|https://img1.doubanio.com/view/photo/s_ratio_poster/public/p1634119717.jpg|8.8
10344754|毒战|未知|剧情,动作,犯罪|https://movie.douban.com/subject/10344754/|https://img3.doubanio.com/view/photo/s_ratio_poster/public/p1924498395.jpg|7.3
10432911|流感|韩国|剧情,惊悚,灾难|https://movie.douban.com/subject/10432911/|https://img1.doubanio.com/view/photo/s_ratio_poster/public/p2029391129.jpg|7.8
10437779|新世界|韩国|剧情,犯罪|https://movie.douban.com/subject/10437779/|https://img3.doubanio.com/view/photo/s_ratio_poster/public/p2507486724.jpg|8.6
10438426|浮城谜事|未知|剧情,犯罪|https://movie.douban.com/subject/10438426/|https://img3.doubanio.com/view/photo/s_ratio_poster/public/p1722545381.jpg|7.3
10439235|创可贴|韩国|剧情,爱情|https://movie.douban.com/subject/10439235/|https://img1.doubanio.com/view/photo/s_ratio_poster/public/p1768410538.jpg|7.3
10441571|朋友2|韩国|动作|https://movie.douban.com/subject/10441571/|https://img3.doubanio.com/view/photo/s_ratio_poster/public/p2154816416.jpg|6.5
10461356|大雄的怀念奶奶|日本|动画|https://movie.douban.com/subject/10461356/|https://img3.doubanio.com/view/photo/s_ratio_poster/public/p2369337420.jpg|9.5
10461367|平安结祈|日本|音乐|https://movie.douban.com/subject/10461367/|https://img3.doubanio.com/view/photo/s_ratio_poster/public/p1407592064.jpg|9.6
cat: Unable to write to output stream.
```

图 8-5　查看文件内容

8.5 Hive 数据分析

Hive 是一个数据仓库基础工具,在 Hadoop 中用于处理结构化数据,它架构在 Hadoop 之上,总归为大数据,使查询和分析变得方便。Hive 还提供简单的 SQL 查询功能,可以将 SQL 语句转换为 MapReduce 任务运行。

1. 创建 Hive 表

(1) 创建数据库

进入 Hive 的命令行工具,创建数据库 douban,命令如下所示。

```
hive> create database douban;
```

(2) 创建表

切换数据库,命令如下所示。

```
hive> use douban;
```

创建 movie 表,命令如下所示。

```
hive> create table movie(
      id STRING,
      title STRING,
      loc STRING,
      cate STRING,
      url STRING,
      cover STRING,
      rate FLOAT
)ROW FORMAT DELIMITED FIELDS TERMINATED BY '|';
```

创建 score 表,命令如下所示。

```
hive> create table score(
      mid STRING,
      uname STRING,
      score INT,
)ROW FORMAT DELIMITED FIELDS TERMINATED BY '|';
```

使用"show tables;"命令查看 douban 数据库中的表,结果如图 8-6 所示。

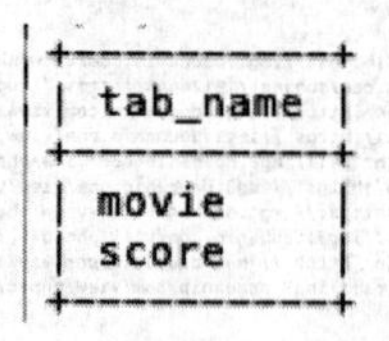

图 8-6 查看库中的表

(3) 加载数据

使用 load 命令将数据集加载到 Hive 表中。加载 movie 表数据集的命令如下所示。

```
load data inpath '/douban/movie/part-m-00000' into table movie;
```

加载 score 表数据集的命令如下所示。

```
load data inpath '/douban/score/part-m-00000' into table score;
```

至此,数据集已经成功导入 Hive,分别查看两个 Hive 表的内容,movie 表的内容如图 8-7 所示,score 表的内容如图 8-8 所示。

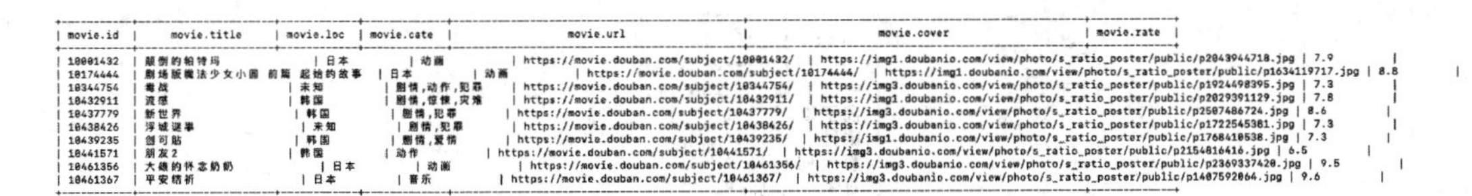

movie.id	movie.title	movie.loc	movie.cate	movie.url	movie.cover	movie.rate
10001432	颠倒的帕特玛	日本	动画	https://movie.douban.com/subject/10001432/	https://img1.doubanio.com/view/photo/s_ratio_poster/public/p2043944718.jpg	7.9
10174444	剧场版魔法少女小圆 前篇 起始的故事	日本	动画	https://movie.douban.com/subject/10174444/	https://img1.doubanio.com/view/photo/s_ratio_poster/public/p1634119717.jpg	8.8
10344754	毒战	未知	剧情,动作,犯罪	https://movie.douban.com/subject/10344754/	https://img3.doubanio.com/view/photo/s_ratio_poster/public/p1924498395.jpg	7.3
10432911	流感	韩国	剧情,惊悚,灾难	https://movie.douban.com/subject/10432911/	https://img1.doubanio.com/view/photo/s_ratio_poster/public/p2029391129.jpg	7.8
10437779	新世界	韩国	剧情,犯罪	https://movie.douban.com/subject/10437779/	https://img3.doubanio.com/view/photo/s_ratio_poster/public/p2507486724.jpg	8.6
10438426	浮城谜事	未知	剧情,犯罪	https://movie.douban.com/subject/10438426/	https://img3.doubanio.com/view/photo/s_ratio_poster/public/p1722545381.jpg	7.3
10439235	创可贴	韩国	剧情,爱情	https://movie.douban.com/subject/10439235/	https://img1.doubanio.com/view/photo/s_ratio_poster/public/p1768410538.jpg	7.3
10441571	朋友 2	韩国	动作	https://movie.douban.com/subject/10441571/	https://img3.doubanio.com/view/photo/s_ratio_poster/public/p2154816416.jpg	6.5
10461356	大雄的怀念奶奶	日本	动画	https://movie.douban.com/subject/10461356/	https://img3.doubanio.com/view/photo/s_ratio_poster/public/p2369337420.jpg	9.5
10461367	平安精祈	日本	音乐	https://movie.douban.com/subject/10461367/	https://img3.doubanio.com/view/photo/s_ratio_poster/public/p1407592064.jpg	9.6

图 8-7 查看 movie 表中的数据

score.mid	score.uname	score.rate
10001432	130451028	4
10001432	134623583	5
10001432	135037922	5
10001432	148126860	4
10001432	26805611	4
10001432	33828431	5
10001432	34213082	4
10001432	38247277	3
10001432	44613271	3
10001432	45498050	3

图 8-8 查看 score 表中的数据

2. 数据分析

接下来使用 SQL 语句对 Hive 表中的数据进行简单的数据分析。

(1) 统计某些类型电影的数量

使用以下 SQL 语句查看类型分别为喜剧、科幻、恐怖、动作的电影的数量。

```
select '喜剧' cate,count( * ) count from movie where cate like '% 喜剧 %'
union select '科幻' cate,count( * ) count from movie where cate like '% 科幻 %'
union select '恐怖' cate,count( * ) count from movie where cate like '% 恐怖 %'
union select '动作' cate,count( * ) count from movie where cate like '% 动作 %';
```

查询结果如图 8-9 所示。

_u3.cate	_u3.count
动作	286
喜剧	348
恐怖	45
科幻	161

图 8-9 根据类别统计电影数量

(2) 根据电影出品地统计电影数量

使用以下 SQL 语句可以根据出品地查看电影数量。

```
select loc,count( * ) from movie group by loc;
```

查询结果如图 8-10 所示。

(3) 统计评分为 5 分的数量排在前十位的电影

使用以下 SQL 语句可以查询好评量排名前十的电影。

```
select id,title,r.count from movie m
JOIN (select mid,count( * ) count from score where rate = 5 GROUP BY mid)r
on m. id = r.mid order by r.count desc
```

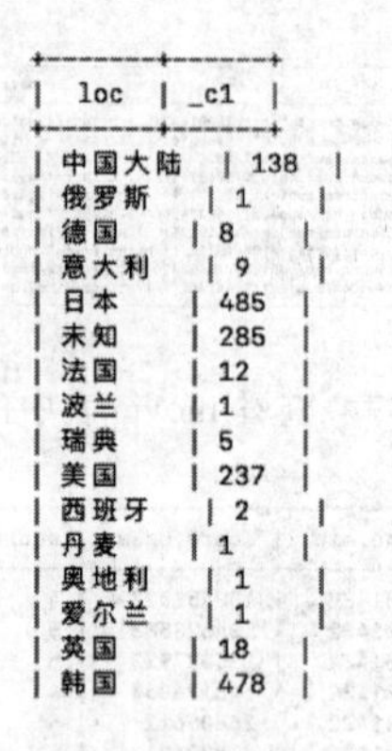

loc	_c1
中国大陆	138
俄罗斯	1
德国	8
意大利	9
日本	485
未知	285
法国	12
波兰	1
瑞典	5
美国	237
西班牙	2
丹麦	1
奥地利	1
爱尔兰	1
英国	18
韩国	478

图 8-10　根据出品地统计电影数量

查询结果如图 8-11 所示。

id	title	r.count
1292722	泰坦尼克号	830
1293182	十二怒汉	810
1291858	鬼子来了	803
1292213	大话西游	802
1292052	肖申克的救赎	802
1292063	美丽人生	796
1295124	辛德勒的名单	796
5912992	熔炉	781
2025894	战长沙	766
1292720	阿甘正传	758

图 8-11　好评量排名前十的电影

8.6　数据挖掘

经过前面几节的操作,本实验已经将实验数据成功导入数据仓库 Hive 中,并通过 SQL 语句从 Hive 中查看了一些数据,接下来将使用 Spark 进行一些更有趣的实验,如利用电影的类别和评分计算电影的余弦相似度,结合用户的评分数据进行基于协同过滤的电影推荐。[44]

由于在将数据导入 Hive 后,原本存储在/douban 下的两个数据文件被移到了 Hive 配置文件中 hive. metastore. warehouse. dir 指定的目录下,为了使用方便,需要利用 Sqoop 将数据重新导入 HDFS,导入方法参考 8.4 节。

1. 基于电影特征计算电影余弦相似度

本实验只需要使用 movie 表,movie 表中的字段如表 8-1 所示。

选用表 8-1 中的电影类别和电影评分字段,计算电影之间的余弦相似度。每个电影都有一个或多个类别,在数据表中形式是“类型 1,类型 2,…”。经统计,该数据集中的电影类别共有 33 种,因此需将 cate 这一列数据展开为 33 列,如果某电影的类别中包含某类别,则将该电影对应的该类别置为 1,否则置为 0。例如,电影《肖申克的救赎》的类别为剧情、犯罪,则将《肖申克的救赎》这一行对应的剧情和犯罪两列置为 1,其余类别对应的列置为 0,如表 8-4 所示。

表 8-4　示例

ID	title	剧情	…	犯罪	恐怖	rate
1292052	肖申克的救赎	1	…	1	0	9.6

Maven 依赖包如下所示。

```
<dependency>
    <groupId>org.apache.spark</groupId>
    <artifactId>spark-core_${scala.version}</artifactId>
    <version>${spark.version}</version>
</dependency>

<dependency>
    <groupId>org.apache.spark</groupId>
    <artifactId>spark-mllib_${scala.version}</artifactId>
    <version>${spark.version}</version>
</dependency>
```

Scala 代码如下所示。

```
import breeze.numerics._
import org.apache.spark.ml.feature.VectorAssembler
import org.apache.spark.sql.{Row, SparkSession}

object Test {
  def main(args: Array[String]): Unit = {
    simMovie("肖申克的救赎")
  }

  def simMovie(movie: String): Unit = {
    //配置 sparksession

    val spark = SparkSession.builder()
    .config("spark.jars","./out/artifacts/SparkTest_jar/SparkTest.jar")
    .master("spark://slave2:8088")
    .appName("douban").getOrCreate()

    //从 HDFS 中读取数据集

    val textFileRdd = spark.sparkContext
    .textFile("hdfs://slave2/douban/movie/part-m-00000").map { x =>
      val tmp = x.split("\\|")
      (tmp(0), tmp(1), tmp(3), tmp(6).toFloat)
    }
    var movie_df = spark.createDataFrame(textFileRdd).toDF("id", "name",
                                                        "cate", "rate")
```

```
val cates = Array("舞台艺术","古装","剧情","历史","情色","武侠","脱口
                  秀","音乐","歌舞","儿童","灾难","科幻","家庭","动
                  作","冒险","运动","纪录片","动画","惊悚","传记","战
                  争","喜剧","悬疑","恐怖","犯罪",“未知","黑色电影",
                  "真人秀","奇幻","爱情","西部","同性","短片")
//将 cate 列扩展为 33 个特征列
cates.foreach { cate =>
  val t = movie_df.withColumn(cate, movie_df.col("cate").contains(cate).
                      cast("int"))
  movie_df = t
}

val features = Array("舞台艺术","古装","剧情","历史","情色","武侠",
                  "脱口秀","音乐","歌舞","儿童","灾难","科幻",
                  "家庭","动作","冒险","运动","纪录片","动画",
                  "惊悚","传记","战争","喜剧","悬疑","恐怖",
                  "犯罪","未知","黑色电影","真人秀","奇幻","爱情",
                  "西部","同性","短片","rate")

//将 33 个类别特征列和评分 rate 列合并为一个向量
val assembler = new VectorAssembler()
  .setInputCols(features)
  .setOutputCol("features")
movie_df = assembler.transform(movie_df)

movie_df = movie_df.select("id","name","cate","features","rate")

//获取电影名为 movie 的电影,movie 是一个变量,由参数传入

var m = movie_df.filter(movie_df("name") === movie).take(1)(0)
var tmp = m.getAs[org.apache.spark.ml.linalg.Vector](3).toDense.values

//计算每部电影与输入的电影之间的余弦距离,排序并取得前十名
movie_df.rdd.sortBy((x =>
   cosine(tmp, x.getAs[org.apache.spark.ml.linalg.Vector](3).toDense.
        values)
  ), false, 1).take(10).foreach(a => println(a(0), a(1), a(2), a(4)))
}
```

```
    //自定义函数,计算两个向量之间的余弦距离,f1 和 f2 分别为两部电影的特征向量
    def cosine(f1: Array[Double], f2: Array[Double]): Double = {
      var sum1 = 0.0
      var sum2 = 0.0
      var sum3 = 0.0
      for (i <- 0 to 33) {
        sum1 = sum1 + f1(i) * f2(i)
        sum2 = sum2 + pow(f1(i), 2)
        sum3 = sum3 + pow(f2(i), 2)
      }
      sum1 / sqrt(sum2 * sum3)
    }
}
```

程序的打包方式在前面的章节中已经有过介绍,此处不再赘述。输出结果如图 8-12 所示。

```
电影名: 肖申克的救赎        类别: 剧情, 犯罪        评分: 9.6
电影名: 教父         类别: 剧情, 犯罪      评分: 9.2
电影名: 美国往事            类别: 剧情, 犯罪        评分: 9.1
电影名: 教父2        类别: 剧情, 犯罪      评分: 9.1
电影名: 完美的世界    类别: 剧情, 犯罪     评分: 9.0
电影名: 偷自行车的人       类别: 剧情, 犯罪        评分: 8.9
电影名: 教父3     类别: 剧情, 犯罪         评分: 8.8
电影名: 四百击       类别: 剧情, 犯罪      评分: 8.7
电影名: 三块广告牌    类别: 剧情, 犯罪     评分: 8.7
电影名: 新世界       类别: 剧情, 犯罪      评分: 8.6
```

图 8-12　推荐结果(一)

由图 8-12 可以看出,计算得到的结果都是与目标电影的类别一致、评分相近的电影。但是在计算电影的相似度时没有利用用户维度的数据,下面的实例将利用 score 表中的用户评分数据基于协同过滤实现电影推荐。

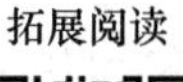

基于用户的协同过滤算法

2. 基于用户评分计算电影相似性

上面的实例将每部电影看作一个向量,然后通过计算向量之间的余弦距离实现了简单的电影推荐,其中没有用到用户维度的数据,实际上可以利用用户的行为数据(如评分)实现电影相似度计算。本书只是基于协同过滤计算出电影相似度,并不会针对某一个体进行推荐,因为前者更加直观,如果读者对协同过滤算法感兴趣,可以自行查阅相关资料进行学习。该算法中,程序会将评分数据转换为一个二维矩阵,如图 8-13 所示,即利用用户的评分来扩展电影的特征,从而组成一个高维的向量,然后进行运算。[45]

用户	电影	分数
小明	流浪地球	4
小明	肖申克的救赎	5
小刚	流浪地球	4
小红	肖申克的救赎	1

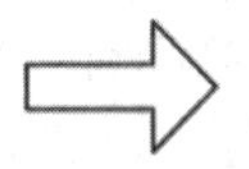

	小明	小刚	小红
流浪地球	4	4	
肖申克的救赎	5		1

图 8-13　矩阵转换

Maven 依赖包如下所示。

```
<dependency>
    <groupId>org.apache.spark</groupId>
    <artifactId>spark-core_${scala.version}</artifactId>
    <version>${spark.version}</version>
</dependency>

<dependency>
    <groupId>org.apache.spark</groupId>
    <artifactId>spark-mllib_${scala.version}</artifactId>
    <version>${spark.version}</version>
</dependency>

<dependency>
    <groupId>org.jblas</groupId>
    <artifactId>jblas</artifactId>
    <version>1.2.4</version>
</dependency>
```

Scala 代码如下所示。

```
import org.apache.spark.ml.feature.StringIndexer
import org.apache.spark.mllib.recommendation.{ALS, Rating}
import org.apache.spark.sql.{DataFrame, SQLContext, SparkSession}
import org.jblas.DoubleMatrix

object Test {
  def main(args: Array[String]): Unit = {
    /* 配置 sparksession */
    val spark = SparkSession.builder().master("spark://slave2:7077")
                .config("spark.jars","./out/artifacts/SparkTest_jar/SparkTest.jar")
                .appName("douban").getOrCreate()

    /* 从 HDFS 中加载数据 */
    val movie_rdd =
    spark.sparkContext.textFile("hdfs://slave2/douban/movie/part-m-00000").map { x =>
      val tmp = x.split("\\|")
      (tmp(0).toInt, tmp(1), tmp(2), tmp(3), tmp(6))
    }
    val score_rdd =
```

```
spark.sparkContext.textFile("hdfs://slave2/douban/score/part-m-00000").map { x =>
  val tmp = x.split("\\|")
  (tmp(1), tmp(0), tmp(2))
}
val movie_df = spark.createDataFrame(movie_rdd).toDF("mid", "name", "loc",
                                                     "cate", "rate")
var score_df = spark.createDataFrame(score_rdd).toDF("uname", "mid", "rate")
/*需要将score表中uname字段的数据全部映射为整型*/
val indexer = new StringIndexer().setInputCol("uname").setOutputCol("uid")
score_df = indexer.fit(score_df).transform(score_df)
/*将score表RDD[ROW]转换为RDD[Rating]类型*/
val r = score_df.rdd
val ratings = r.map { row =>
   Rating(row.getDouble(3).toInt, row.getString(1).toInt, row.getString
        (2).toDouble)
}
/*训练模型,使用Spark MLlib包中的ALS*/
val model = ALS.train(ratings, 40, 20, 0.01)
/*目标电影名*/
val movieName = "教父"
val itemId = movie_df.filter(movie_df("name") === movieName).take(1)(0).
                                                   getInt(0)
/*获取目标电影向量*/
val itemFactor = model.productFeatures.lookup(itemId).head
val itemVector = new DoubleMatrix(itemFactor)
/*计算相似度*/
val sims = model.productFeatures.map { case (id, factor) =>
  val factorVector = new DoubleMatrix(factor)
  val sim = cosineSimilarity(factorVector, itemVector)
  (id, sim)
}
/*取前十名*/
val sortedSims = sims.top(10)(Ordering.by[(Int, Double), Double] { case
                             (id, similarity) => similarity })
/*输出结果*/
sortedSims.foreach { row =>
  val tmp = movie_rdd.filter { x =>
    x._1.toString.equals(row._1.toString)
  }
```

```
      val name = tmp.first()._2
      val cates = tmp.first()._4
      val rate = tmp.first()._5
      println(s"电影名:$name \t 类别:$cates \t 评分:$rate")
    }
  }
  /*计算余弦相似度,和上面的实例中的方法不同,这里利用 DoubleMatrix 实现了矩阵
    运算*/
  def cosineSimilarity(vec1: DoubleMatrix, vec2: DoubleMatrix): Double = {
    vec1.dot(vec2) / (vec1.norm2() * vec2.norm2())
  }

}
```

输出结果如图 8-14 所示。

```
教父      类别: 剧情, 犯罪 评分: 9.2
爆裂鼓手 类别: 剧情, 音乐 评分: 8.6
真爱至上 类别: 剧情, 喜剧, 爱情评分: 8.5
这个杀手不太冷类别: 剧情, 动作, 犯罪评分: 9.4
蝴蝶效应 类别: 剧情, 科幻, 悬疑, 惊悚评分: 8.7
大话西游之月光宝盒类别: 喜剧, 爱情, 奇幻, 冒险 评分: 8.9
爱在午夜降临前   类别: 剧情, 爱情 评分: 8.8
勇往直前 类别: 剧情, 传记, 灾难   评分: 8.4
催眠大师 类别: 剧情, 悬疑, 惊悚   评分: 7.6
重庆森林 类别: 剧情, 爱情         评分: 8.7
```

图 8-14　推荐结果(二)

从图 8-14 中可以看出,此方法的效果没有第一种方法的效果好,主要是因为用户评分的数据量太少,数据质量不佳。第一种方法所用的数据都是豆瓣官方给出的数据,且不需要大数据量,如果数据足够的话,第二种方法的效果要比第一种方法的效果更好。在工业界的实际应用场景中,基于协同过滤的推荐算法被广泛应用。

参考文献

[1] 刘鹏. 大数据实验手册[M]. 北京：电子工业出版社，2017.
[2] Apache Software Foundation. HBase 官方文档中文版[EB/OL].[2019-09-18]. http://abloz. com/hbase/book. html.
[3] Dimiduk N,Kurana A. HBase 实战[M]. 谢磊,译. 北京:人民邮电出版社,2013.
[4] Redis 翻译团队. Redis 中文官方网站[EB/OL]. [2019-03-16]. http://www. redis. cn.
[5] MongoDB 中文社区[EB/OL]. [2019-09-08]. http://mongoing. com.
[6] 肖鹏. Neo4j 中文手册[EB/OL]. [2019-09-11]. http://neo4j. com. cn/public/docs/index. html.
[7] Cypher 中文文档[EB/OL]. [2019-10-01]. http://neo4j. com. cn/public/cypher/default. html.
[8] 董西成. 大数据技术体系详解:原理、架构与实战[M]. 北京:机械工业出版社,2018.
[9] 林子雨. 大数据基础编程、实验和案例教程[M]. 北京:清华大学出版社,2017.
[10] Maven Getting Started Guide[EB/OL]. [2019-11-27]. http://maven. apache. org/guides/getting-started/index. html.
[11] 苦涩咖啡. MapReducer 二次排序——原理[EB/OL]. [2018-06-14]. http://blog. chinaunix. net/uid-26111972-id-5786277. html.
[12] 夏延. Mapreduce 实例——自定义输出格式[EB/OL]. [2018-09-07]. https://www. cnblogs. com/aishanyishi/p/10304887. html.
[13] 夏俊鸾. Spark 大数据处理技术[M]. 北京:电子工业出版社，2015.
[14] 林大贵. Hadoop + Spark 大数据巨量分析与机器学习整合开发实战[M]. 北京:清华大学出版社，2017.
[15] jackiehff. Spark 2. 2. x 中文官方参考文档[EB/OL].(2018-01-19)[2019-08-24]. https://spark-reference-doc-cn. readthedocs. io/zh_CN/latest/.
[16] 高彦杰. Saprk 大数据处理:技术、应用与性能优化[M]. 北京:机械工业出版社，2015.
[17] 王家林,徐香玉. Spark 大数据实例开发教程 [M]. 北京:机械工业出版社，2016.
[18] Armbrust M, Xin R S, Lian C, et al. Spark sql: Relational data processing in spark[C]//Proceedings of the 2015 ACM SIGMOD international conference on management of data. ACM，2015：1383-1394.
[19] Zaharia M, Chowdhury M, Franklin M J, et al. Spark: Cluster computing with working sets[J]. HotCloud，2010，10(10-10)：95.
[20] Meng X, Bradley J, Yavuz B, et al. Mllib: Machine learning in apache spark[J]. The Journal of Machine Learning Research，2016，17(1)：1235-1241.

[21] Abbasi M A. Lean Apache Spark 2 [M]. Birmingham:Packt Publishing，2017.

[22] Karau H，Konwinski A，Wendell P，et al. Learning spark：lightning-fast big data analysis[M]. Sebastopol：O'Reilly Media，Inc.，2015.

[23] Apache Software Foundation. Storm 官方网站[EB/OL]. [2019-03-25]. http://storm.apache.org.

[24] Apache Software Foundation. Flume 官方网站[EB/OL]. [2019-03-27]. http://flume.apache.org.

[25] Apache Software Foundation. Kafka 官方网站[EB/OL]. [2019-03-27]. http://kafka.apache.org.

[26] Apache Software Foundation. Spark 官方网站[EB/OL]. [2019-03-27]. http://spark.apache.org.

[27] 肖立涛. Spark Streaming 实时流式大数据处理实战[M]. 北京:机械工业出版社,2019.

[28] Apache Software Foundation. Flink 官方网站[EB/OL]. [2019-04-16]. http://flink.apache.org.

[29] Hive 四种数据导入方式介绍 [EB/OL].（2018-05-21）[2019-02-16]. https://www.cnblogs.com/shujuxiong/p/9067651.html.

[30] 卡普廖洛,万普勒,卢森格林. Hive 编程指南[M]. 曹坤,译. 北京:人民邮电出版社,2013.

[31] 刘博宇. Druid 在滴滴应用实践及平台化建设 [EB/OL].（2018-06-06）[2019-02-06]. https://yq.aliyun.com/articles/600128? utm_content=m_1000000412.

[32] 欧阳辰,张海雷,高振源,等. Druid 实时大数据分析原理与实践[M]. 北京:电子工业出版社,2017.

[33] Pydruid: A python connector for Druid[EB/OL].（2015-10-02）[2019-09-25]. https://github.com/druid-io/pydruid.

[34] Druid 官方教程[EB/OL]. [2019-09-05]. http://druid.io/docs/latest/tutorials/tutorial-batch.html.

[35] pydrill 0.3.4 documentation[EB/OL]. [2019-09-24]. https://pydrill.readthedocs.io/en/latest/readme.html# sample-usage.

[36] Drill 官方教程[EB/OL]. [2019-09-24]. https://drill.apache.org/docs/querying-json-files/.

[37] Kylin 官方网站[EB/OL]. [2019-02-27]. http://kylin.apache.org/cn/.

[38] Apache Kylin 核心团队. Apache Kylin 权威指南[M]. 北京:机械工业出版社，2017.

[39] 蒋守壮. 基于 Apache Kylin 构建大数据分析平台[M]. 北京:清华大学出版社，2017.

[40] 倪彬彬. 浅谈大数据可视化[J]. 福建电脑,2018,34(11):101-102.

[41] D3 官方网站[EB/OL]. [2019-05-22]. http://d3js.org/.

[42] 肖明魁. Python 在数据可视化中的应用[J]. 电脑知识与技术,2018,14(32):267-269.

[43] max_hello. 用 sqoop 将 mysql 的数据导入到 hive 表中[EB/OL].（2018-09-27）[2019-09-25]. https://www.cnblogs.com/xuyou551/p/7998846.html.

[44] 晁岳攀. 使用 Spark MLlib 给豆瓣用户推荐电影[EB/OL].（2015-11-30）[2019-09-25]. https://colobu.com/2015/11/30/movie-recommendation-for-douban-users-by-spark-mllib/.

[45] NeilZhang. 推荐系统——协同过滤原理与实现[EB/OL].（2018-11-03）[2019-09-26]. https://www.cnblogs.com/NeilZhang/p/9900537.html.